COMMON GROUND

This book is dedicated
to the memory and legacy
of

BLACK ELK

May the Tree Bloom Again

COMMON GROUND

ECO-HOLISM & NATIVE AMERICAN
PHILOSOPHY

ROY C. DUDGEON, Ph. D.

PITCH BLACK PUBLICATIONS

COMMON GROUND:

Eco-holism and Native American Philosophy

All contents Copyright 2008 by Roy C. Dudgeon

ISBN: 978-1-4357-1738-1
Library of Congress Control Number: 2008905222

PITCH BLACK PUBLICATIONS
Winnipeg, Manitoba, Canada
http://stores.lulu.com/pitchblackpublications

Printed by Lulu Enterprises,
860 Aviation Parkway, Suite 300,
Morrisville, NC, 27560
USA.

Contents

Contents

INTRODUCTION

It is well understood that the only decent future for us who live in America now is through a rediscovery of our environment. We need to establish a right relationship with the land and its resources; otherwise, the destruction of the Indian will be followed by the destruction of nature; and in the destruction of nature will follow the destruction of ourselves. The Indians, in a sense, knew this all along...Perhaps now, after hundreds of years of ignoring their wisdom, we may learn from the Indians (McLuhan 1971:1-2).

Because of their uniquely keen perception of the natural order, Amerindians have felt from their first meeting with Europeans that these people were coming here out of a pressing need to relearn their own relationship with other humans and the rest of creation. Even if the first Americans foresaw that the coming of the Europeans would produce a shock both overwhelming and catastrophic for them and for the land, they could also imagine what would come after: Whites, horrified by the socio-ecological mess they had made of the continent, would one day turn to Native people and ask their help in re-establishing order...(Sioui 1992:111).

THE PRESENT WORK is inspired by two streams of thought. The first is ecological philosophy, which shall be referred to as *eco-holism* in what follows[1]. Eco-holism is a comparatively new philosophy in the Western world, which represents a radical departure from the modern philosophy from which our current technologies and ways of life arose. It has arisen in recent decades in response to empirical evidence that there is a fundamental and growing contradiction between the aims and activities which Western culture has traditionally pursued and the possibility of its own perpetuation. This contradiction is suggested by the multitude of interlocking ecological problems manifesting themselves around the world; such as deforestation, soil erosion and species extinctions, nuclear and chemical pollution, overpopulation and overconsumption, ozone

---

[1] For a fairly complete discussion of the varieties of eco-holism and many of the key issues which it raises see Merchant (1992). For a succinct discussion of three of the better known varieties of eco-holist thought–including deep ecology, eco-feminism and social ecology–as well as their shared premises, see Dudgeon (2008:11-33).

depletion, acid rain and climate change, to name a few of the most obvious[2] . In light of these facts, eco-holism suggests that we can no longer continue to act upon the world as a mere commodity, without regard for the fact that we are totally dependent upon nature in order to survive. For if we do, our powerful modern technology is fully capable of subverting the very conditions which sustain our lives.

The second stream of thought which inspires the present work is Native American philosophy, as it is expressed in their oratory, literature and ways of life. Not surprisingly, as eco-holism has continued to develop and extend its influence, growing numbers of people have commented upon the similarities in outlook which eco-holism and Native American philosophies share (Bird-David 1993; Booth and Jacobs 1990; Callicott 1982, 1994; Harrod 2000; Hughes 1983, 1991, Knudtson and Suzuki 1992; Reed 1991). Many contemporary Native Americans, including scholars, also recognize the affinity, as shall become evident in the chapters to follow, and as is illustrated by the introductory quotation from Sioui, a contemporary scholar of Huron descent. Sioui's sentiments also closely mirror those of McLuhan, a Euro-American, who clearly intended to tie her early compilation of excerpts from Native American literature to ecological themes.

Anyone who has studied both areas in any detail cannot help but notice the many similarities between the two streams of thought, which should not be ignored. As Mircea Eliade suggests regarding indigenous philosophies in general, "Western philosophy is dangerously close to 'provincializing' itself...by jealously isolating itself in its own tradition" (1954:x). In the present case,

---

[2] For an excellent discussion of the interlocking nature of ecological problems and the possible synergistic effects between them–which he refers to as "the nemesis effect," see Bright (2001).

such a view is not only unjust–in that it ignores the fact that Native America seems to have already arrived at many of the views which eco-holism is only now rediscovering–but also unwise. For in the present context it is the Western way of conceiving of and acting upon the world which is itself proving to be problematic. Isolating ourselves exclusively within that tradition, then, will not help us to see our way out of these difficulties. It is more likely to perpetuate them. Through comparison with the views of others, however, we may learn where our own patterns of thought and action have gone awry, for as Gregory Bateson once suggested with regards to our current ecological problems, "[i]t is possible that some of the most disparate epistemologies which human culture has generated may give us clues as to how we should proceed" (Bateson and Bateson 1988:136).

There are, of course, those who would disagree, particularly when it comes to using Native American thought as an inspiration for developing more ecological philosophies, attitudes and styles of life in the Western world. Calvin Martin, for example, once concluded that "There can be no salvation in the Indian's traditional conception of Nature for the troubled environmentalist. Some day, perhaps, he will realize that he must look to someone else other than the American Indian for realistic spiritual inspiration" (1978:188). Martin is not alone in questioning the appropriateness of turning to Aboriginal peoples and their philosophies as an inspiration for the development of more ecological ways of thinking and living in the West, and other dissenting opinions shall be considered where appropriate in what follows[3].

One of the central purposes of the present work, however, shall

---

[3] Callicott (1982) offers a detailed reply to such objections from an eco-holist point of view, while Cordova (1997:31-44) provides a critique of Callicott's methodology and assumptions from the perspective of an Apache scholar.

be to offer a model for describing the similarities in outlook which eco-holism and Native American philosophies seem to share which I hope may help to expand the growing dialogue between the two streams of thought, and begin to illustrate what Western culture may learn from Aboriginal philosophies. Examining the differences between the ideologies of the colonizing cultures and those which were colonized–in order to construct this comparative model–will further illustrate the many similarities eco-holism shares with Aboriginal philosophies as well.

I will be approaching this task from an holistic perspective, and this in two senses. First, these two broad cultural types–the Native American and the Occidental–shall be viewed within the larger context of cultural evolution which is provided by an anthropological perspective. The second and primary sense in which the present work may be termed "holistic," however, is derived from eco-holism, and especially from the works of Bateson–this being a central emphasis upon the study of *patterns of relationship*, rather than upon "wholes[4]."

Human beliefs change, and with them change also our patterns of relating to one another, and also to the larger world. Neither are these three areas–the ideological, the social, and the ecological relationships–in any way separable from on another. Rather, in the view of the present work, they are simply different ways of describing aspects of the same interrelational context[5].

---

[4] For a more detailed discussion of the difference between relationalism and "whole-ism" see Dudgeon (2008:1-11).

[5] Those familiar with recent anthropological theory will note a similarity here to Marvin Harris' distinction between superstructure, structure and infrastructure. I have chosen not to use this terminology due to certain theoretical differences with Harris' position–especially its materialism, determinism and positivism, and its emphasis upon what Harris calls "infrastructural determinism"–which shall be discussed in more detail below. For the best account of his position see Harris (1979).

Just as meaningful or ideological relations are a *part of* the larger context of social relations, so too, social relationships are subsumed by the larger context of the ecological relationships in which society participates. Yet each is simply a more *inclusive* description of the *same* context, and should not be considered in isolation from one another. Instead, picture them as a series of three spheres, nested one within the other, with each larger sphere being a *precondition* for the existence of those within it.

Just as all ecological relationships are not social, but do provide the *enabling conditions* for there being social relationships, so too, all social relationships among living beings are not *meaningful* in the sense in which human social relationships often are. Yet it is *within* such patterns of social relations that meaningful patterns of relationship are given the possibility to arise–through the emergence of symbolic patterns within them. Still, these patterns of meaning remain inseparable from the larger ecological context in which such social and behavioral relations take place. It is these patterns of relationship–both as enacted in daily life, and as illustrated and sanctioned by philosophy and religion–which shall provide the focus in the following.

Such a pattern of relationship–of smaller patterns nested within larger ones, and of the relationships between the various more inclusive scales–shall be termed a "holarchy" in what follows. My use of this metaphor is derived from two sources. The first is Bateson's discussion of "orders of recursiveness" (1979:218), while the second is John Lame Deer's description of Lakota cosmology as a pattern of "circles within circles within circles, with no beginning and no end" (1972:100). Thus, the holarchy metaphor may be derived from either of the two schools of thought which inspire the present work[6].

The central methodology of the present work—of comparing relational symbols with one another, and with the social and ecological patterns of relationship of various societies—is also based upon Bateson, and especially upon his discussion of the methodology of *abduction* (1979:153-55). Bateson described abduction as a method of comparing *patterns of relationship* and their *symmetry or asymmetry*. Abduction is, for example, the methodology used in traditional comparative anatomy to sort individual creatures into species, genera, families and the like, based upon similarities and differences in their physical characteristics, or phenotypes. As shall be discussed below, Bateson himself often used this methodology to compare and contrast the patterns of relationship in the social and ecological spheres[7].

What the present work adds to this approach is an emphasis upon the importance of *relational symbolism*, which is also derived from my study of both ecological philosophy and Native American traditions. After all, ecology defines itself as the study of *patterns of interrelationship* in the natural world. Since humanity is a part of nature, and since symbol systems are a part of human life, symbol systems are also a part of nature. As shall be discussed below, relationalism is also one of the most important premises shared by the eco-holist and Native American schools of thought. Thus, the way in which *ideal* patterns of relationship are represented philosophically—that is, *relational symbols*—becomes my central focus in

[6] The term "holarchy" was originally suggested by Koestler (1978:34), though I am not a follower of Koestler's theoretical orientation, nor of his interpretation of the meaning of the term.

[7] His discussion of abduction is also closely connected to his definition of explanation as "the mapping of description onto tautology" (1979: 85-93, Bateson and Bateson 1988:90) and with his discussion of "syllogisms in grass" (1988:26-30). Further discussions of the methodology of abduction may be found in: Dudgeon (2008:113-15) and Harries-Jones (1995:177-80).

the various philosophies considered below. This is because relational symbols appear to be important vehicles through which the lifestyles consistent with any particular philosophy are propagated and maintained[8].

Philosophy, therefore, will not be conceived of as an activity which is peculiar to certain professional academics in the Western world. Instead, as Hester and McPherson suggest, "Philosophy is a thoughtful interaction with the world" (1997:9). Furthermore:

> Indigenous philosophy must be treated with the same respect as European philosophy, Indigenous people must be recognized as the central stake holders, inheritors, and developers of their philosophy; the philosophy must be recognized as a complete way of life, containing ethical, social, political, epistemological, as well as metaphysical elements (1997:9).

Consequently, the present work shall not rely too heavily upon the secondhand interpretations of Native American religions and philosophies provided by Western ethnographers, except as an aid in placing them within a broader historical and ecological context. In light of my training as an anthropologist, however, I also realize that some level of ethnographic detail and context is still expected in works dealing with nonwestern peoples, which is why it shall still be provided where it seems appropriate through-out. Even so, the present work should be read, first and foremost, as a philosophical inquiry, or as an exercise in cross-cultural philosophy, rather than as an ethnographic work.

Further, I shall also attempt as much as possible to base my interpretations of Native American philosophies upon Native American literature, or upon sources which have been dictated or written by Native Americans themselves–that is, upon the primary

---

[8] While my study of Native American philosophies is an important inspiration for my emphasis upon relational symbolism throughout the present work, it is important to point out that this emphasis is not, in itself, an "indigenous methodology," which would be a more appropriate description of the efforts of contemporary indgienous scholars.

literature.

While the earliest written accounts of their words date back at least four centuries, the twentieth century witnessed an explosion of firsthand Native literature. There thus appears to be little excuse in the present day for not turning to Native American literatures themselves for confirmation of one's interpretations of Native American philosophies.

Indeed, the literature in question forms a very self-consistent whole, despite its internal variety–not unlike a school of thought in Western philosophy, or a research paradigm in the social sciences. Further, turning to their own explanations and justifications in this way can only lead to a deeper and richer understanding of the world view and ethos to which their works give expression than that which would be attainable by relying upon secondhand sources alone. Their views may then be directly compared to the philosophical traditions of the Western world on an equal footing, since access to the two points of view shall be gained in the same manner–through an inspection of their words and works. Both may then be compared to the philosophy of eco-holism. Yet as Cordova, a contemporary Apache scholar, cautions:

> The person of the West exists, for the West, as the standard by which all other persons are measured...Western thinkers have organized the world on a hierarchical scale wherein they reserve for themselves the highest rung. At each level below them they place all other peoples according to how closely they resemble the lifestyles and value systems of the West (1997:35).

Such an ethnocentric attitude is obviously the antithesis of what is being attempted here. Thus, in the following discussion, the ideology of Western culture shall not be assumed to be superior to that of others, nor given a special dispensation when it comes to its veracity. Rather, Western philosophy will be placed upon the same "level" as those of other cultures, ages and peoples. This is certainly a prerequisite if an honest comparison is to be made. Indeed, since part of the central task of the present work is to understand the common ways in which eco-holist and Native American philosophies differ from dominant Western philosophies, as well as their common critiques of the same, traditional Western philosophies shall often be compared to the former two and found to be wanting. For as Clifford Geertz once described the central aim of his interpretive anthropology, the task is "to make available to us answers that others, guarding other sheep in other valleys, have given, and thus to include them in the consultable record of what man has said" (1973:30). Such an attitude certainly seems more conducive to the task at hand.

The primary goal of the present work, therefore, is that of cross-cultural comparison, especially of patterns of belief and behavior at the ideological, social and ecological levels. Due to the fact that my primary goals are ethnological and comparative, rather than ethnographic and descriptive, I shall also often speak in generalist, rather than particularist terms about "the West" and "the Native" as ideal types. What should be kept in mind throughout, however, so as to avoid reducing either category to mere stereotypes, is that both categories are characterized by a great deal of internal diversity. The same may be said of the patterns of behavior of the individuals within any society. The latter does not change the fact that there are still general patterns of behavior and belief to be observed at the cultural level, however, and the same remains true at the level of the broader ideal types or cultural groupings which form my subject matter. In other words, both cultures and cultural types may still be usefully characterized in terms of their dominant characteristics, or in terms of the patterns of belief and behavior which are most predominate, or which best characterize them, and individual cultures, or even individual persons, may still serve as examples of those general patterns.

One of the most well known primary sources on Native American philosophies remains the teachings of Black Elk, the renowned Oglala Lakota holy man. His teachings are preserved in two books: *The Sacred Pipe* and *Black Elk Speaks*. The latter in particular has become a classic in the field, being widely quoted by both ecologists and Native Americans alike, and held up as a main example or illustration of Native American philosophy. My own reading of *Black Elk Speaks* was one of the central inspirations which lead me to research this topic, and to approach it in the manner in which I have–through the extensive use of Native American autobiographies, oratory and writings. It also provides us with an important example of issues concerning the "authenticity" of such sources.

Because many of the earliest sources of Native American views are *translations* from Native languages into English, questions of "authenticity" often arise for those with a "textual" or "postmodern" emphasis (in the literary sense). The skill and biases of the translator, they point out, are likely to have had an impact upon the shape of the final English narrative. Because of the great differences in the grammatical structures of Native American languages and English, furthermore, only an expert in both languages could capture all of the subtle nuances of meaning, and transform them accurately into English. Indeed, only such an expert could fully understand the extent to which some concepts and expressions are simply *untranslatable*, or lose certain connotations or associations when taken out of their original linguistic and cultural context.

Even so, these are questions which apply to a greater or lesser extent to *all* translations from one language to another, and do not invalidate these sources–lest we reintroduce a double standard through the back door. Many of the classics of academia, for example, were not written in English either. Yet we do not throw out the philosophical study of the works of Plato and Aristotle merely because they were originally written in ancient Greek. Neither do we eschew the study of Socrates' philosophy simply because we have only Plato's account of it, nor those of the Pre-Socratics because we know them only from scattered quotations in the works of others.

To be fair, neither should we disregard early sources of Native American philosophy simply because they were originally narrated in Lakota or Cree, nor because we may have only a few excerpts from any particular thinker. Rather, *all* such translations should be recognized as consisting of one person giving an account of what another person has written or said, and should properly be considered to be collaborations, regardless of their provenience[9].

The over-all consistency of Native American literature, however, and the persistence of certain key themes from the earliest accounts of their views to the present day, can serve as a corrective for the failings which may be inherent in particular sources. The present work shall also draw upon many sources which were

---

[9]. This is not to imply that successful translation from Native American languages into English is not a more difficult undertaking than performing the same task from one European language into another, for the former does involve "communication between very different cultures. The problems...are not simply those inherent in the necessity to translate...into English, as would be the case in a situation where the witnesses where Francophone. French and English cultures, although different, trace common historical roots and share a worldview. The Gitksan and Wet'suwet'en world view is of a qualitatively different order" (Wa and Ukwa 1992:22). The understanding of differences of the latter type is, of necessity, a central task of the present work.

originally written or narrated in English–whose authors thus did their own translation–which should provide a balance for any shortcomings which may be inherent in sources which did require translation.

The authenticity of *Black Elk Speaks* has been questioned in recent years for reasons closely related to these. Due to the publication of the original interview notes from which the book was prepared, John G. Neihardt, who collected Black Elk's narrative and edited it into a book, is now known to have taken some liberties when presenting Black Elk's words to the public. Though the transcripts were certainly polished and somewhat transformed in the process, a comparison reveals that they generally retained the spirit of what was said, if not the exact wording. As Raymond DeMallie observed in *The Sixth Grandfather*, for example, in which these transcripts were first published:

> Neihardt was an extraordinarily faithful spokesman for Black Elk; what he wrote was an interpretation of Black Elk's life, but not one that was embellished in any way. Instead, he tried to write what he thought the old man himself would have expressed (1985:51).

One must also remember that what was being polished and edited was already a *translation* in any case–that is, an account of what someone else had said[10]. As DeMallie observes, even during the translation process, there is a sense in which "Neihardt was already 'writing' Black Elk's story by rephrasing his words in English," (1985:51). This again reminds us of the *collaborative* nature of all works in translation. Yet while *Black Elk Speaks* does not represent the actual words of Black Elk, one must also remember

---

[10] Black Elk's own son, Ben, acted as the principal translator of his Lakota narrative, with Neihardt assisting in rendering the account into English. The account was then written out by Neihardt's daughter, Enid, and eventually typed into a "final" draft which Neihardt used during the preparation of the manuscript copy of *Black Elk Speaks*.

that the original interview transcripts are not Black Elk's own words either, but rather a reconstruction of his original Lakota narrative in English. Where he felt it was necessary, Neihardt also added historical details from other sources in order to fill in background information which was not included in the transcript.

There seem to be two ways of dealing with this situation. One is to throw out *Black Elk Speaks* and make use only of the original transcripts provided by *The Sixth Grandfather*, because they are "closer" to the original narrative. This will be the position of many Western academics, who are most concerned with the fact that the book does not preserve Black Elks "actual words"[11]. This is not, however, the position which shall be taken in what follows.

As has already been stated, Black Elk's own words *no longer exist*, since his original Lakota narrative was, unfortunately, *never recorded in that language*. So while the transcripts are certainly closer to a verbatim account, and preserve some details which were omitted in *Black Elk Speaks, both* are translations. And though the precise wording may have been altered in the latter, the philosophy presented in the two seems to be largely consistent. It also seems consistent with the teachings which were later preserved in *The Sacred Pipe*, which were recorded by Joseph Epes Brown, who seems to have made less use of poetic license in translating Black Elk's words. To what extent Neihardt's later additions and changes to the transcripts were based upon memories of unrecorded con-

---

[11] See, for example, the work of Julian Rice (1991). Rice suggests that the transcripts published in *The Sixth Grandfather* contain "a more authentic voice," whereas *Black Elk Speaks*, "may perhaps be relegated to the ranks of nineteenth [sic] century curios, reflecting white misconceptions of Indians" (1991:13-14). While Rice spends much of his time analyzing Neihardt's other writings in order, it would appear, to invalidate the interpretation he presents in *Black Elk Speaks* through association with opinions expressed elsewhere, his work, like *The Sixth Grandfather*, is worth consulting by all serious students of Black Elk's philosophy.

versations between himself and Black Elk is impossible to know, as is the extent of Black Elk's own participation in the process.

I shall, therefore, make use of the second option, and will continue to quote from *Black Elk Speaks*. There are many reasons for doing so. The first is that many passages are already well known, being widely quoted by ecologists and Native Americans alike. As Vine Deloria, Jr. points out in his Introduction to a recent edition of *Black Elk Speaks*, "the book has become a North American bible of all tribes" (Black Elk 1988:xiii). Thus, if there are doubts as to whether it perfectly preserves the historical beliefs of the Lakota in general, or of Black Elk in particular, there can be no doubt that it is representative of the beliefs of many Native Americans in the present day, for it has been widely embraced as a religious classic. This fact alone should be all the justification necessary for continuing to quote from it. Sioui (1992:8), for example–who is a Huron (or Wendat) scholar writing in French, quotes from a French translation of the English translation of Black Elk's philosophy (of which I have read an English translation)–and yet he does not seem troubled by issues of "authenticity."

Its continued acceptance even in the face of the original transcripts also seems to reveal a different emphasis upon its importance on the part of Native American scholars. This is a concern not so much with its *accuracy* or "authenticity" as an historical document, so much as with the continuing *truth* of the teachings which it preserves–surely a much more philosophical issue.

Nor was Black Elk's individual reputation as a holy man and teacher based solely upon these books. As Frank Fools Crow points out, among his own Lakota people, "My uncle, the renowned Black Elk, has earned a place above all of the other Teton holy men. We all hold him the highest" (1979:53). Though Fools Crow had never read his books, he had learned many things from Black Elk first hand, and went on to become a holy man and ceremonial leader of the Lakota tribes in his own right. Fools Crow also continued the tradition which his uncle had initiated by sharing his own teachings and visions in two books which were given to Thomas E. Mails (Fools Crow 1979, 1991). Again it is the religious or philosophical significance of Black Elk's teachings which is central to his reputation, rather than questions of historical accuracy. And from this perspective, the "authenticity" argument could even be seen as an attempt to dismiss as irrelevant what is considered by many to be a sacred document.

Finally, we must also consider what Black Elk's own intentions were in sharing his stories and teachings. After all, it would appear that Black Elk himself wanted not only to preserve something of the history and ways of his people, but felt that it was important to pass on the great vision with which he was gifted in his youth as well. As a child, Black Elk was sick for twelve days, during which time he lay unconscious and near death in his parents' teepee. During this time he experienced a prophetic vision, in which he was taken to a meeting with the Six Grandfathers, who represented the powers associated with the six sacred directions of the world. In the vision, they revealed to him many sacred teachings, as well as sharing with him a vision of the future of his people.

In the symbolism of the vision Black Elk saw his people climbing a path which made four ascents. These he interpreted as representing four generations of his people. The first two ascents were to be quite good, with his people walking the path in a sacred manner, and retaining their sacred teachings, rituals and ways of life. This sacred path was represented by the hoop of the nation, which was also representative of the unity of all things. At the centre of the hoop grew a sacred flowering tree, representing the prosperity of the people, and of the entire Earth. But in the third ascent the hoop was to be broken, the tree was to whither, and the people were to fall upon difficult times. As the transcript records his words:

> His dream has been coming true. The first and the second ascents were both good and the third is to be a fearful thing and perhaps we're in that time now–something is going to happen...In Black Elk's days he has seen the second generation and in the third he thinks something fearful is going to happen...from there on every man has his own vision and his own rules. The fourth ascent will be terrible...There were lots of sick children–all pale and it looked like a dying nation. They showed me a circle village and all the people were very poor and there you could hear the wail of women and also men. Some of them were dying and some were dead (DeMallie 1985:126-28)[12].

Yet the Six Grandfathers gifted Black Elk with the powers of a healer as well. When he grew to adulthood he was to use the power given by his vision in order to become a shaman or medicine man, and to cure the ills of his people. He was given power not only to cure the sicknesses of individuals, but to restore the hoop as well, and to make the flowering tree bloom once more at its centre. By doing so he would set the people back upon the sacred path which their ancestors had walked, and restore harmony to the creation. As he states, "In the vision I was representing the earth and everything was giving me power. I was given power so that all creatures on earth would be happy" (DeMallie 1985:133).

---

[12] This passage also illustrates one of the deficiencies of the transcript as a translation, particularly in the earlier portion–the constant shifting of tense, from second person descriptions of what Black Elk said, to presumed first person utterances, which Neihardt polished into a first person account throughout *Black Elk Speaks*. For the complete account of the vision see Black Elk (1988:20-47), and DeMallie (1985:111-42). DeMallie's comparison of the two is also useful (1985:93-99).

This would happen at the end of the fourth ascent, which thus represented both the closing of a cycle, and the dawning of a new age. As Black Elk states, "Depending upon the sacred stick we shall walk and it will be with us always. From this we will raise our children and under the flowering stick we will communicate with our relatives–beast and bird–as one people. This is the center of the life of the nation." It is easy to see why his vision is often read as a metaphor for the ecological crisis which we are now facing, and perhaps of its resolution, particularly when he adds that "This tree never had a chance to bloom because the white man came" (DeMallie 1985:130).

Thus, Black Elk felt that he had failed in the mission which the Six Grandfathers had bestowed upon him. Because of the influence of the Western world, the sacred hoop was broken, and the tree representing the Indian spirit, and the sacred traditions of his people, never bloomed as it should have. Black Elk even gave up his practice as a medicine man, thus turning away from the mission and power of his vision, and joined the Catholic Church. The desire to preserve his vision, however, and to somehow restore the sacred hoop and cause the tree to flower as it was meant to never left him.

He found a new vehicle in Neihardt, whom he immediately recognized as a kindred spirit. By giving the vision to Neihardt, he was not only preserving it for the future, but seems to have felt that he was passing on his spiritual burden as well, along with the power which he had been given to make the tree bloom. As Black Elk states at one point:

> The more I talk about these things the more I think of old times, and it makes me feel sad, but I hope that we can make the tree bloom for your children and for mine...it was hopeless it seems before I saw you, but here you came. Somehow the spirits have made you come to revive the tree that

never bloomed (DeMallie 1985:44).

So it seems that Neihardt had Black Elk's blessing when it came to interpreting his vision to the world, for in a sense he was the inheritor of its power. It would now be Neihardt's place to bring its message and teachings to the world, and in this sense the power to make the tree bloom, and to restore harmony in the creation, was now his. Indeed, such a passing on of medicine teachings, or of the right to tell particular stories, has always been quite common in cultures with an oral tradition, and seems perfectly legitimate from such a perspective.

Black Elk had also adopted Neihardt into the Lakota tribe, naming him "Flaming Rainbow." In his great vision, the flaming rainbow had stood above the doorway to the lodge in which Black Elk had received the teachings of the Six Grandfathers. The flaming rainbow thus represented the portal through which truth and wisdom were gained, and through which the power of the vision was brought to the world.

The significance of the name which Black Elk chose thus tells us quite a lot about his perceptions of both Neihardt, his adopted nephew, and of the act of passing on his vision to another man. Indeed, it would seem that in transforming his narrative into a more accessible and persuasive form, Neihardt was only fulfilling the obligation which Black Elk himself had placed upon him–to bring his vision to as many people as possible so that the tree might bloom again.

Thus, I shall continue to quote from *Black Elk Speaks* in what follows, for I am concerned not only with its authenticity as an historical document, but also with the truth of its teachings, and I shall attempt to walk the fine line between these two perspectives[13].

The purpose of the present work, then, is not only to understand Native American philosophies as they were historically, but to understand what eco-holism might learn from them–both past and present. Indeed, partly because it was one of my own central inspirations for writing the present work, the symbolism of Black Elk's vision has also been embodied into the structure of the text itself–four sections of four chapters, with an introduction and conclusion. These represent the four sacred directions, as well as Earth and Sky or, in totality, "The Six Grandfathers" and their teachings.

The central theme of balance has also been integrated into the book's structure in a variety of ways. For example, introduction and conclusion represent opposite but complementary terms; the beginning of my own (or the reader's) intellectual journey in a state of relative ignorance, and the end of that journey in a state of (one hopes) relative enlightenment. The first two main sections of the manuscript also represent the past, and the dualism or antagonism, between Western and Native, while the latter two sections represent the present, and the potential for a "balancing" of indigenous and Western views (and thus the closing of the circle, or the potential ending of the cycle of antagonism between the two cultural groups). Finally, the four central sections also move back and forth between a consideration of theoretical or philosophical issues (Sections I and III), and the social, ecological or practical issues relevant to the time period discussed[14].

---

[13] For those most concerned with questions of "authenticity," corresponding passages from the interview transcripts shall also be provided–where such exist–in the footnotes for these quotations. In this way interested readers may judge for themselves as to whether their meaning was preserved, at least from one English translation to another.

[14] I will leave it to readers to find additional examples.

Structuring the manuscript in this way not only provided a certain mental discipline to my thinking throughout the process of preparing it, but also makes the structure of the argument a relational symbol in its own right. In other words, the symbolism embodied in the work's structure not only serves as an example of the work's methodology and themes, but serves as an additional means through which those issues are communicated on another symbolic level as well.

Let us turn, in the first section, to a more detailed account of the relational symbolism which we find represented in Native American philosophies and narratives, followed by an abductive comparison with those of the Western world. Section two shall then discuss the manner in which the relational symbolism of each cultural type was reflected in their patterns of social and ecological relationship. In other words, it shall compare the relational symbolism of the ideological sphere to the patterns of relationship in the social and ecological spheres.

Section three shall then return to more theoretical concerns, or to a direct comparison of Native American and eco-holist philosophies. This shall include a discussion of their common differences from, and critiques of, the dominant ideological, social and ecological patterns of relationship of the contemporary Western world. Finally, section four shall again return to a discussion of the theory of practice, though in the contemporary context. More specifically, it shall discuss ecological and indigenous issues in the context of contemporary trends towards the globalization of economic organization, especially by elaborating upon the eco-holist critique of contemporary capitalist organization, technology and philosophy. This section shall also attempt to illustrate the manner in which eco-holism is both consistent with–and supportive of–contemporary Native American aspirations towards greater political autonomy, economic self-sufficiency and control over the management of resources on their traditional lands–as well as the two traditions' potential to be supportive of one another in practice.

I

THE CIRCLE WHICH UNITES
& THE LINE WHICH DIVIDES

1. THE MOTHER GODDESS: BALANCE AS AN IDEAL

> You have noticed that everything an Indian does is in a circle, and that is because the Power of the World always works in circles, and everything tries to be round...Everything the Power of the World does is done in a circle. The sky is round, and I have heard that the earth is round, like a ball, and so are all the stars. The wind, in its greatest power, whirls. The birds make their nests in circles, for theirs is the same religion as ours. The sun comes forth and goes down again in a circle. The moon does the same, and both are round. Even the seasons form a great circle in their changing, and always come back to where they were. The life of a man is a circle from childhood to childhood, and so it is in everything where power moves. Our teepees were round like the nests of birds, and these were always set in a circle, the nation's hoop, a nest of many nests where the Great Spirit meant for us to hatch our children (Black Elk 1988:194-96)[15].

IRONICALLY, the earliest prehistoric evidence which suggests that elements of philosophies similar to those of Native America may

---

[15] This is one of the most famous passages in *Black Elk Speaks*, and is widely quoted in contemporary Native American literature (Cajete 2000:259; Sioui 1992:9). A closely corresponding section of the interview notes reads:
"You will notice that everything the Indian does is in a circle. Everything that they do is the power from the sacred hoop, but you see today that this house is not in a circle. It is a square. It is not the way we should live...The power won't work in anything but circles. Everything is now too square. The sacred hoop is vanishing among the people. We get even tents that are square and live in them. Even the birds and their nests are round. You take the bird's eggs and put them in a square nest and the mother bird just wont stay there. We Indians are relative-like with the birds. Everything tries to be round—the world is round. We Indians have been put here to be like the wilds and we cooperate with them. Their eggs of generations are in the sacred hoop to hatch out. Now the white man has taken away our nest and put us in a box and here they ask us to hatch our children, but we cannot do it. We are vanishing in this box" (DeMallie 1985:290-91).

have developed comes not from the Americas, but rather from the Paleolithic hunting cultures of Europe. This evidence consists of two types of art–Paleolithic cave paintings; which include many depictions of animals, the hunt, and some figures which are thought to be shamans, as well as in a contemporaneous form of sculpture; the "Venus" figurines. These little footless and often faceless female figurines, often with exaggerated breasts and hips, were found throughout much of southern Europe, and are often interpreted as early material evidence that Goddess worship was one of the earliest expressions of religious belief in human culture. As Joseph Campbell states it: "The female figurines are the earliest example of the 'graven image' that we possess, and were, apparently, the first objects of worship of the species *Homo sapiens*" (1969:325).

Where the cave art is thought to have been related to male hunting magic and rites, the Venus figurines were often found in the actual shelters, where families lived, and were thus usually associated with the domestic scene. Whether these figures were direct objects of worship is, of course, something of which one can never be certain. But the fact that they were highly stylized pieces, while contemporaneous forms of sculpture depicting animals were highly realistic, suggests to many a religious significance of some kind for these figures. Several also appeared to have been set up in shrines when found, which further suggests a status as religious objects of some type (Campbell 1969:313). According to Campbell:

> These suggest that the obvious analogy of women's life-giving and nourishing powers with those of the earth must already have led man to associate fertile womanhood with an idea of the motherhood of nature (1969:139).

This interpretation, while speculative, would give the association of nature and the Earth with the feminine an ancient history.

It is also a common view in traditional Native American religions, if not one of their defining features. According to Black Elk, "the growing power is rooting in mystery like the night, and reaches lightward. Seeds sprout in the darkness of the ground before they know the summer and the day. In the night of the womb the spirit quickens into flesh" (1988:209)[16]. Or again, when speaking of a woman's first menstruation, he states that "the change in her is a sacred thing, for now she will be as Mother Earth and will be able to bear children" (1971:116). The creative power inherent in the female form is considered to be analogous to the creative and nurturing power immanent in nature. Indeed, within this type of *animistic* cosmology, the entire universe is thought to be alive–animated by this same *immanent* power.

The figure of the Mother Goddess has often been presented as a giver of life in other ways as well; as the source from which food springs, or as a patroness of the hunt. The White Buffalo Woman of the Sioux or Lakota people, who was the bringer of their Sacred Pipe, and with it their religion and way of life, was both a Goddess and a buffalo in human form. She thus provided a further reason for them to respect and honour the buffalo, their principle source of food and provision, as well as the Earth which nourished them, for it was from the buffalo people that their patroness had come[17].

Where for hunting peoples the Goddess is often the source and protectress of game, for many horticultural peoples She is embodied in the food plants themselves. A Penobscot legend, for example, tells of how First Mother was approached during a time of hunger, when game had become scarce due to a growing

---

[16] Compare this to the quote from *The Sacred Pipe* which follows.

[17] For various versions of the legend of the White Buffalo Woman and the coming of the Sacred Pipe, see Black Elk (1971: 3-9,1988:3-5), DeMallie (1985:283-85), Fools Crow (1990:142-44), Lame Deer (1972:240-44) and Erdoes and Ortiz (1984:47-52).

population and, consequently, over-hunting. She informs the people that they must kill her, drag her body over the ground until the flesh is torn from the body, and then bury the bones. Seven months later corn sprouts from this spot, and the hunger of the people is relieved, while tobacco springs from her bones[18]. The plant's fruit–corn–was First Mothers flesh, given so that the people might live and flourish. The people are thus subsisting upon the body of a sacrificed Goddess, whose nourishing power is immanent within the food they eat, and the Earth from which it grows, while the tobacco, her second gift, was central to worship and ritual (Erdoes and Ortiz 1984:12-13).

Besides being representative of a sacred power immanent in nature, the Mother Goddess also represents the ultimate unity behind all duality, for just as she is the governess of life, so also is she the governess of its necessary corollary–death. Describing a similar myth to the one above, Campbell states that:

> at the moment of the sacrifice, when death came into the world and with it the flow of time, there occurred also a separation of the sexes; so that with death there came the possibility of procreation and birth. The pairs-of-opposites, thus, of male and female, death and birth...came into the world, together with food, at the end of the Mythological Age (1984:78).

She is "both the facilitator of transformations and the enclosing, protecting, and embracing governess of the process" (1984:80). Both death and life flow from her; she is the unity which lies behind the dualities of life. Just as plants and animals spring from the womb of the Earth, later to die and return their substance to her, only to be reborn once more; so also in creating life, death also comes into being, and from death, ultimately, life arises again. Or as Chief Dan George once expressed it; "Life and death–a song

---

[18] This is also a possible reference to the fact that horticulture was invented by women.

without an ending" (1982:54)[19].

The celestial symbol for this type of cyclicity is the phases of the moon in its waxing and waning. As Black Elk notes, "the moon represents a person and, also, all things, for everything created waxes and wanes, lives and dies" (1971:71). Just as the moon passes from light to darkness to light, a cyclicity which unites the opposites within a larger unity, so also the fertility of women follows the same cycle with the waxing and waning of their "moon time," and the fertility of the Earth waxes with the coming of spring and wanes with the coming of fall, ever returning to where it has been.

Indeed, this symbolism of the unity or complementarity of opposites renders Campbell's designation of it as the mythology of the "Goddess" something of a misnomer. For the symbol of the Goddess is seldom emphasized in such a way as to totally exclude the importance of the male principle. Rather, as Campbell himself admits, within this type of mythology "divinity could be represented as well under the feminine as the masculine form, the qualifying form itself being merely the mask of an ultimately unqualified principle, beyond, yet inhabiting all names and forms" (1964:13). This is more in keeping with Native American religions, where both Sky and Earth, the Father above and the Mother below, are honoured together, as complementary aspects of the Great Mystery. For as Charles Alexander Eastman[20] relates:

> the Indian no more worshipped the Sun than the Christian adores the Cross. The Sun and the Earth, by an obvious parable, holding scarcely more

---

[19] For a detailed discussion of Native American attitudes towards death from the perspective of a Lakota scholar, which also contrasts such ideas with the Christian or secular attitudes of the Western world, see Deloria (1994:165-83).

[20] Eastman was a mixed blood Sioux who grew up on the plains as a forager, but was later educated at one of the early American boarding schools. He went on to study medicine, and was working as a doctor at the Pine Ridge Agency during the Wounded Knee massacre.

of poetic metaphor than of scientific truth, were in his view the parents of all organic life. From the Sun, as the universal father, proceeds the quickening principle in nature, and in the patient and fruitful womb of our mother, the Earth, are hidden embryos of plants and men. Therefore our reverence and love for them was really an imaginative extension of our love for our immediate parents (1980:13-14).

Thus it is the power of the Sky–the life giving warmth of the sun and the nourishing rains–which fertilize the Earth, whose hidden powers then give shape and substance to the living things born from it[21]. Both are equally necessary to life, for and as Cajete suggests, "Native communities embody and harmonize the duality of maleness and femaleness, for these complementary relation-ships ensured the survival of the group" (2000:94). After all, the life of a human infant is only made possible through the analogous coming together of its mother and father in sexual relations. Thus, as Black Elk asks, "Is not the sky a father and the earth a mother, and are not all living things with feet or wings or roots their children" (1988:3)[22]? Yet as Cajete suggests, while both powers are acknowledged, Native philosophies do contrast with the Occidental in the sense that:

> Life and divinity in the Native view are experienced through the feminine, life-giving principle of the Earth. Native perceptions of the

[21] Indeed, from an ecological perspective, all of the *energy* in the biosphere *does* come from the sky–in the form of solar energy converted by green plants, which is passed up the food chain to higher trophic levels–while all of the *material* in the biosphere is derived from the Earth.

[22] A similar reference to the Sky as a Father and the Earth as a Mother is found in Neihardt's stenographic notes. The passage describes the vision which Black Elk saw when he was taken to the center of the world (Harney Peak in the Black Hills) by the Six Grandfathers:

"And while I stood there I saw more than I can tell and I understood more than I saw; for I was seeing in a sacred manner the shapes of all things in the spirit, and the shape of all shapes as they must live together like one being. And I saw that the sacred hoop of my people was one of many hoops that made one circle, wide as daylight and starlight, and in the center grew one mighty flowering tree to shelter all the children of one mother and one father. And I saw that it was holy" (DeMallie 1985: 97).

land and its qualities are primarily Earth centered rather than sky centered as perceived in many Indo-European cultures (2000:185).

Ultimately, however, the creative power has both male and female aspects. This symbolism also includes both that which is "beyond," or transcendent, as well as that which "inhabits," or the immanent. Indeed, Black Elk also emphasizes the importance of these other complementary aspects of the Great Mystery–the immanent and the transcendent–in the following way:

> We should understand well that all things are the works of the Great Spirit. We should know that He is within all things: the trees, the grasses, the rivers, the mountains, and all the four legged animals, and the winged peoples; and even more important, we should understand that He is also above all these things and peoples (1971:xx).

This symbolism of the unity or complementarity of opposites is often the theme of Native American legends as well, where it is represented as the coming into balance of opposite tendencies or extremes. As Jordan Wheeler suggests, "[t]he victory in the Aboriginal story is when harmony can be achieved between the character and his/her environment" (1992:39)[23]. While there are many different ways in which this ideal is represented, the symbolism is sometimes particularly striking.

One such example is the Blood-Piegan legend which describes "how men and women got together." The story begins shortly after Old Man has finished creating the world. Yet while Old Man felt that "he had done everything else well," he had made one mistake in creating the world, in that he had originally put men and women in separate villages, and had created them so that "men and women did everything exactly the same way." Both knew how to drive buffalo off of cliffs, and how to butcher the

---

[23] Wheeler is a former editor of *Weetamah*, Manitoba, Canada's Aboriginal newspaper at this time.

animals for meat, but this was the only food they knew. After a time, however, each group began to develop unique skills of its own. The men invented the bow and arrow, and learned how to hunt for other types of food. The women learned how to tan hides, as well as to make and decorate clothing and tipis.

Then, one day, Old Man realized the mistake he had made in putting men and women in separate camps, saying to himself, "Men and women are different from each other, and these different things must be made to unite so that there will be more people. I must make men mate with women. I will put some pleasure, some good feeling into it; otherwise the men won't be keen to do what is necessary." He then decided that he must first set an example himself, which the other men could follow, so he set off to take a look at the women's camp, and then returned to tell the other men of all the good things he had seen there. Unlike the men, they lived in fine homes of tanned hides. And where the men wore only "a few pelts around their loins," the women wore finely made leather clothing. When the rest of the men heard of all "the useful and beautiful things" the women made, they immediately decided that they should "go over there and get together with these different human beings."

Meanwhile, the chief of the women's village had seen the tracks which Old Man left, and trailed them to the village of the men. She was astonished, and hurried back to her camp with the news, exclaiming to everyone, "Oh, sisters, these beings live very well, better than us. They have a thing shooting sharp sticks, and with these they kill many kinds of game–food that we don't have. They are never hungry."

As the story continues, the two camps try several times to get together. When the men approach the first time, the women are appalled at what they see. The men are dirty and smelly, with matted hair and only a dirty pelt for clothing. Forgetting of the good things their leader has told them, the women throw rocks at the men and drive them off. At this the men become angry, and even Old Man states that, "It was no mistake putting these creatures far away from us. Women are dangerous. I shouldn't have created them."

Later, however, both groups changed their minds, having had time to think again about the many useful skills the others have. So Old Man bathed himself and dressed in a fine outfit of clothing which he had taken from the women's village[24]. When the men approach this time he tells them to hang back, while he goes into the village alone to speak to the women. Yet a scout from the women's camp has already seen them coming, and has run to tell the others where they are butchering buffalo. The woman chief decides that they should remain just as they are–dressed in their worst clothing, covered in blood from their work, and with their flint skinning knives still in their hands. When Old Man sees the woman chief like this he exclaims, "She's dressed in rags covered with blood. She stinks. I want nothing to do with a creature like this. And those other women are just like her." He than walked away, with all of the other men following him.

But then the women realized their mistake, and the next time they take the initiative. They dress in their best clothing, decorated in fine porcupine quill designs, and don bone and shell necklaces and bracelets. They wash and comb their hair and paint their faces. Then they set off for the village of the men.

Here the men have again changed their minds, and can think of nothing but the women. Then a sentry comes running into camp

---

[24] An example of theft by the creator.

telling of the women's approach. Immediately the men rush to the river to bathe. They clean and arrange their hair, decorate it with feathers, paint their faces, scent themselves with cedar incense, and put on their best fur clothing. Finally, men and women meet on terms which are pleasing to both, since each is attempting to please the other.

The woman chief and Old Man were immediately attracted to one another. Remembering his original reason for wanting to bring men and women together, Old Man invites the woman chief on a walk alone. When they have some privacy, he says to her, "Let's try one thing that has never been tried before." The woman chief replies, "I always like to try out new, useful things." They then decided that it might be best to lie down while trying this.

Afterwards, they both agreed that this was the most pleasurable thing they had ever experienced, and rushed to tell all the others about their wonderful discovery. But when they reached the village they found that all of the other men and women were nowhere to be seen, since "they didn't need to be told about this new thing; they had already found out." "Then the women quilled and tanned for the men. Then the men hunted for the women. Then there was love. Then there was happiness. Then there was marriage. Then there were children."

And thus it was only by realizing that their differences were complementary, rather than separate or opposed, that men and women achieved a harmonious relationship with one another, and the human race was allowed to perpetuate itself (Erdoes and Ortiz 1984:41-5). The fact that, first, the Creator made a mistake, second, that he was not omnipotent (lacking the skill as a man to make fine clothing for example) and, finally, that he taught the people how to have sex through leading by example, all provide interesting contrasts with the Old Testament version of the creation, which shall be discussed below.

Another excellent example of the themes of balance and harmony in Native American oral literature is the Swampy Cree legend of "The Battle of the Four Winds." This story begins when a Cree man, whose wife has just given birth to twins, has a prophetic vision, "in which the Indian saw the spirit of fire, who told him that he would be visited by the four winds of the world on the day his twin children became two years old." Thus the theme of dualities is made explicit from the beginning. When two years had passed, the man had made the elaborate preparations which the spirit of fire had described to him, and with his wife and children sheltered in an underground cellar beneath his specially constructed lodge, he awaited the arrival of the four winds.

The north wind—the wind of cold and of winter—was the first to arrive. Appearing in the guise of a woman, she entered the lodge and blew her cold breath upon the Indian until he was near freezing. She was followed shortly by the wind of darkness—the wind which comes from the place of the setting sun in the west—and even the light of his fire was extinguished when it entered. They were shortly joined, however, by the warm wind of the south and the summer—who entered in the form of a man—and the wind of light from the east—the place of the dawn.

The Indian then witnessed a great battle between these opposing forces, with the winds of warmth and of cold sending up a great gray mist as they came together, and with the lodge flashing in daylight and darkness as the east and west winds clashed. But after "fighting back valiantly," the north wind realized that "her resistance was futile; she continued to thaw and diminish in size." The south wind does not seek to destroy her, however. Instead, he states that "you will never again dominate the world. From now on I will replace you part of the year." Similarly, the winds of light and darkness eventually fight themselves to a stalemate, and "Realizing that one could not defeat the other, they made an agree-ment to share the world. They decided to take equal turns visiting the earth" (Ray and Stevens 1971:62-3).

In this way, harmony was achieved between the female wind of the north and the male wind of the south, between summer and winter, and between the forces of light and darkness. As Luther Standing Bear states of the beliefs of his own people, "the philosophical ideal of the Lakota was harmony, and the most powerful symbol of that was peace" (1978:40). This is certainly the case with the Cree legend given above as well, in which peace and harmony are achieved simultaneously.

It is this type of symbolism which Wheeler seems to be refer-ring to when he states that in Native American stories, "Good and evil are traded in for the Aboriginal concern with balance and harmony" (1992:38). For neither extreme is triumphant. Instead, winter is balanced with summer, light with darkness, and male with female, with each pair becoming complementary to one another, rather than separate and opposed. And where the north and the night had dominated, this imbalance is corrected. Both the state of imbalance and the animosity which had existed between each pair of opposites are overcome, and a state of peace and harmony is achieved within the creation.

So in saying that good and evil are not Aboriginal concepts, Wheeler is not suggesting that they have no conception of right and wrong, but that their ethics represents their ideals differently. It is believed that exclusively pursuing either of the extremes is imbalanced or disharmonious, and that the ideal lies not at either of the extremes, but at the point of balance between them. As Black Elk observes, "the truth comes into the world with two faces. One is sad with suffering, and the other laughs; but it is the same face, laughing or weeping" (1988:188)[25]. The opposites are seen as complementary aspects of a larger whole, for as he adds–when describing the symbolism of the two thongs used to tie a Sun Dancer to the center pole, which are actually a single piece of hide–"although there seem to be two thongs, the two are really only one; it is only the ignorant person who sees many where there is really only one" (1971:95).

This sentiment is echoed by Lame Deer when he states that "You are isolated, but you know that you are part of the Great Spirit, united with all living things" (1972:172). Thus, both individuality and the relationship to the larger unity which lies behind this apparent diversity are recognized and considered, with neither being focused upon to the exclusion of the other. It is this type of balanced understanding which the Native shaman or holy man always seeks, as Lame Deer relates with reference to Christian symbolism:

> A medicine man shouldn't be a saint...He should be able to sink as low as a bug, or soar as high as an eagle. Unless he can experience both, he is no good as a medicine man...You can't be so stuck up, so inhuman that you want to be pure...You have to be God and the devil, both of them. Being a good medicine man...means experiencing life in all its phases (1972:68).

To achieve such a balanced understanding one must look at things from many different perspectives, or points of view, and learn from whatever wisdom lies in each. As Lame Deer states it, "[w]e are forever looking at things from different angles," and since both the positive and the negative are aspects of the world, one should consider what each can teach one, for "[i]t is all part of the

---

[25] Compare this to the passage from *The Sacred Pipe* which follows.

same thing–nature, which is neither sad nor glad; it just is" (1972:190). The medicine man–as healer, religious leader, and philosopher–simply looks to the world around him and observes that–as an Iroquois legend concerning Hiawatha the Unifier expresses it–"it is the law of the universe that happiness alternates with sorrow, life with death, prosperity with hardship, harmony with disharmony" (Erdoes and Ortiz 1984:196). So while harmony remains the ideal, even disharmony offers its own lessons, and its own type of wisdom[26]. For as Lame Deer observes:

> Nature, the Great Spirit–they are not perfect. The world couldn't stand that perfection. The spirit has a good side and a bad side. Sometimes the bad side gives me more knowledge than the good side (1972:68).

---

[26] As Cajete (2000:217) suggests, the cosmic principle of Chaos is often represented in Native American mythology by humorous stories of the Trickster figure who, even while being a sacred clown who is constantly doing things incorrectly and making mistakes, is also a creator.

2. NATURAL ARCHETYPES: THE CIRCLE & THE MEDICINE WHEEL

the Indian's symbol is the circle, the hoop. Nature wants things to be round. The bodies of human beings and animals have no corners. With us the circle stands for the togetherness of people...The camp in which every tipi had its place was also a ring. The tipi was a ring in which people sat in a circle and all the families in the village were in turn circles within a larger circle, part of the larger hoop which was the seven campfires of the Sioux, representing one nation. The nation was only part of the universe...circles within circles within circles, with no beginning and no end. To us this is beautiful and fitting, symbol and reality at the same time, expressing the harmony of life and nature (Lame Deer 1972:100).

AS SEEMS EVIDENT from the above, the symbol of the circle is inclusive, rather than exclusive. It brings things together within progressively larger contexts rather than separating them from one another. The family is a circle within the circle of the tribe's camp, which is a circle within the hoop of the people, which is in turn contained within the larger sphere of nature. Readers should also note the similarity to my own description of the relations between the ideological, the social and the ecological spheres given above, which shall become a central theme in Section III, when ecological and Native American philosophies will be abductively compared to one another, and contrasted with modern Occidental philosophy. For in such a view, people conceive of themselves as *participants* within nature. Indeed, as Cajete points out, "nature was the primary model for Native community. Plants, animals, natural phenomena, earth, sun, moon, and cosmos were used as symbols and models for emulating" (2000:104). Rather than emphasizing a division between the cultural or social sphere and the larger context of nature, they see themselves as merely another part of the natural world, no better or worse than any other part of the Great Spirit's creation.

In the social sphere this leads to a very egalitarian philosophy, consistent with the ideal of reciprocity among the members of the band or tribe which most Native American groups traditionally practiced Interestingly, the symbol of the circle also represents egality in the mythology and symbolism of the Western world. One need look no farther than the popular legends of King Arthur for an example of this. For the knights met, of course, at a round table; a symbol of the fact that they were champions all, and met with their king as equals. This was not, however, the usual practice of kings historically.

Yet because of the inclusive tendency of Native American thought, their egalitarian views did not end even with humanity, but were extended to include all living things within an even greater circle of reciprocal relations. Indeed, Cajete describes this type of "mutual reciprocity" not only as a relationship with other humans, but as "a give-and-take relationship with the natural world," which he suggests is "one of the oldest ecological principles practiced by Indigenous peoples all over the world" (2000:79). Or as Jenny Leading Cloud, a Lakota elder observes:

> we Indians think of the earth and the whole universe as a never ending circle, and in this circle man is just another animal. The buffalo and the coyote are our brothers; the birds, our cousins. Even the tiniest ant, even a louse, even the smallest flower you can find–they are all relatives. We end our prayers with the words *mitakuye oyasin*–all my relations–and that includes everything that grows, crawls, runs, creeps, hops, and flies on this continent (Erdoes and Ortiz 1984:5).

This participatory philosophy, when combined with the idea that the Great Spirit is immanent within the creation, leads to an ideal of respect for all of the things of the natural world, for all are considered to be kin. As Chief Dan George encapsulates the views of his people, "The earth and everything it contained was a gift of See-see-am...and the way to thank this great spirit was to use his gifts with respect" (1974:37). The world, and everything within it–including other people–must be treated with respect, since all things are seen as manifestations of the Great Spirit itself. To abuse them, exploit them, or belittle them is to scoff at the sacred power itself, which is immanent within them. For as Lame Deer relates, "To us a man is what nature, or his dreams, make him. We accept him for what he wants to be. That's up to him" (1972:139). Such an attitude is also extended to other animals, for as Pete Catches observes "all animals have power, because the Great Spirit dwells in all of them, even a tiny ant, a butterfly, a tree, a rock" (as cited in Lame Deer 1972:116). All living things are afforded ethical regard, and indeed, "Everything is alive" (as cited in Lame Deer 1972:212)[27]. As Luther Standing Bear articulates these views:

> Wakan Tanka [the Great Spirit] breathed life and motion into all things, both visible and invisible. He was over all, through all, and in all, and great as was the sun, and good as was the earth, the greatness and goodness of the Big Holy were not surpassed. The Lakota could look at nothing without at the same time looking at Wakan Tanka, and he could not, if he wished, evade His presence, for it pervaded all things and filled all space (1978:197).

Thus, all things within the Great Spirit's creation are considered to be equal, and all are related to one another by the sacred power which moves within and through all. This sacred power is the Great Spirit itself. Creation, therefore, is not understood as a singular event which took place long ago in a period of six days (as in the Old Testament mythology). Rather, from an animistic perspective, creation is a *continuous* or ongoing process. In other words, it is the immanent presence of the Creator *within* things which makes

---

[27] The speaker is Leonard Crow Dog, another Lakota holy man.

their existence possible. As Robinson describes Mi'kmaq[28] understandings of the Creator, for example:

> Christian dogma or doctrine cannot be adequately translated into a system
> compatible with the Mi'kmaw way of thinking...the notion that God
> oversees the universe from afar is inimical to the 'true Mi'kmaw mind,'
> which perceives the process of creation as unfolding and as something in
> which the Creator continues to participate fully (2005:35)[29].

Consequently, since all things are manifestations of the Great
Spirit's power, which flows through them, and is seen as making
their lives possible on an ongoing basis, all must be respected if the
Great Spirit is to be respected. As Standing Bear continues, this
attitude of respect for other living things imposed certain duties
upon people:

> The animal had rights–the right to man's protection, the right to live, the
> right to multiply, the right to freedom, and the right to man's
> indebtedness...This concept of life and its relations was humanizing and
> gave the Lakota an abiding love. It filled his being with the joy and
> mystery of living; it gave him reverence for all life; it made a place for all
> things in the scheme of existence with equal importance for all. The
> Lakota could despise no creature, for all were of one blood, made by the
> same hand, and filled with the essence of the Great Mystery (1978:193).

It is from this viewpoint, which Cajete describes as "[t]he philosophical ideal of ethical *participation* with nature" (2000:83, original
emphasis), that the common practice of ritually apologizing to

---

[28] The Mi'kmaq are an Algonquian speaking people (like the Cree, who become
a central ethnographic example below) whose traditional territory consists of
what is now the Canadian Maritime provinces. They are often referred to in
the ethnographic literature as "Micmac," though the appropriate plural is
"Mi'kmaq," and the singular "Mi'kmaw," following the contemporary Smith-
Francis orthographic system (Robinson 2005:2).

[29] Robinson's reference is to a recorded narrative from a Mi'kmaw "tradition-
alist" (whose identity is not revealed for reasons of privacy) concerning the
philosophical implications of the typically Algonquian structure of his
language, which is primarily verb based. As he states, "A concept of 'god'
in...Mi'kmaw language is...not stationary. In fact, the words...that we have
for 'god' are all verbs...and that's an indicator that 'god' is not someone who is
sitting up in the sky who oversees all things" (Robinson 2005:35).

game taken in the hunt arose, which was common to many tribes. Though killing and eating other beings is necessary for survival, the hunter must assure the animals who give up their lives for his benefit that he and his family will one day die and return their bodies to the Earth, to provide the substance from which other lives may grow, including the kin of the animal slain. As Roger Jones, an elder to his own Shawanaga First Nation once stated:

> when we take life, we offer our tobacco. If we are going to take a deer, we ask that deer if we can take its life so that we can sustain our own life and provide food and clothing for our family. And when that deer gives us that life, we again give our tobacco and say *Miigwetch*. Thank you very much for giving your life for us. And that is the same with all of the things around us[30].

Indeed, as Wooden Leg of the Cheyenne made the same point almost a century before, this type of view extended to trees and grasses, as well as to all other things. As he stated it; "Whatever one of such growths may be destroyed by some good Indian, his act is done in sadness and with a prayer for forgiveness because of his necessity, the same as we were taught to do in killing animals for food or skins" (1931:374). For all of the creation is sacred, and is thus worthy of being treated with respect. One should not take more than one needs, and greed for possessions is seen as

---

[30] Jones belonged to the Algonquian speaking Ojibway or Aniishinabe peoples who inhabit the areas North and West of Lake Superior. He was speaking to Canada's Royal Commission on Aboriginal Peoples' (RCAP) in Sudbury, Ontario on June 1, 1993. The RCAP was an official, arm's length tribunal commissioned by the government of Canada to assess the current state of Canada's aboriginal peoples, and to offer recommendations upon improving their condition. The CD-ROM version of the RCAP's extensive five volume final report (and other writings) from which I quote contains no page numbers. This quote appeared in V. 1, Part III, "Building the Foundation of a Renewed Relationship," Chapter 15.6 "The Land That Supports Us."

destructive, wasteful, and imbalanced[31].

The common symbol of the medicine wheel--a cross inscribed within a circle--unites the ideal of balance and the teachings of the circle within a single unified figure. The Four Directions are represented as points on the circle, and are united by two lines from north to south, and from east to west. These two lines are often described as two paths, as they were by Black Elk, who described them as "the good red road" and the black "road of difficulties." These represented a balanced or harmonious path running east to west--following the pattern of celestial motion--and an imbalanced path from north to south, respectively. In this symbol, "we see that everything leads into, or returns to, the center; and this center which is here, but which we know is really everywhere, is *Wakan Tanka*" (1971:89-90). Thus the ideal--the sacred--is represented by the center; by the point midway between east and west, north and south; the point of balance where the harmonious path and the inharmonious path cross. And this center represents the Great Spirit; the Great Mystery; the Big Holy--the unity which lies behind and within the apparent diversity of all things, and the ideal of parts and wholes moving together in harmony.

The symbolism of the medicine wheel and the sacred centre was often integral to various ritual performances as well. An altar constructed of Earth, and in the shape of a medicine wheel, was often central to larger ceremonies and rites. This is true of the Sun Dance, for example, which is a ritual common to several of the First Nations which inhabited the Great Plains, including the Lakota and Cheyenne. The Sun Dance is considered by the Lakota to be their

---

[31] For a more detailed discussion of the manner in which this philosophy was incorporated into Native American rituals among various First Nations, and of the way in which it served to regulate Native ecological relationships, see Hughes (1983).

most sacred ceremony. It is an offering of themselves to *Wakan Tanka* in self-sacrifice; as a prayer for the well-being of all things, as well as being an important means of seeking sacred visions.

The Medicine Wheel symbolism is also embodied in the dance itself. Here the sacred center is represented by the *can-wakan*, or sacred cottonwood tree, which forms the center of the circular lodge in which the people dance. During the performance of the dance, pledgers are (most commonly) pierced on each breast and tied to the center pole with rawhide thongs, symbolizing their connection to *Wakan Tanka*, and thus to all things of the Earth as well. The pledgers then dance, pulling back upon the thongs, until they can tear them from their skin, and free themselves[32]. This act is thus symbolically analogous to a rebirth or a renewal, with the dancers breaking free of the direct connection to Mother Earth–as from their human mother at birth–and regaining the limited separateness which we all enjoy as individuals. This symbol of the circle with the sacred tree at its centre was central to Black Elk's vision as well, just as it was to the rituals of his people, and as he suggested:

> In the old days when we were a strong and happy people, all our power came to us from the sacred hoop of the nation, and so long as the hoop was unbroken, the people flourished. The flowering tree was the living center of the hoop, and the circle of the four quarters nourished it. The east gave

---

[32] Because the Sun Dance involved flesh offerings and the piercing of the dancers, it was widely misunderstood and condemned by Christian missionaries and many others. Consequently the Sun Dance was explicitly banned by the American government for many years, in spite of their constitution guaranteeing religious freedom. It continued to be practiced in secret throughout this period, however, and has been openly revived in the last few decades. For Lakota accounts of the Sun Dance see Black Elk (1971:67-100); Fools Crow (1990:114-38); Lame Deer (1972:187-202); McGaa (1990:85-9)6; Standing Bear (1975:113-22); for an account of the Cree Sun Dance (or Thirst Dance) see Dion (1979:36-42); for a less detailed account of the Crow Sun Dance in which a woman acted as the sponsor see Pretty Shield (1974:208-12); and for an ethnographer's description of the Cheyenne Sun Dance see Hoebel (1978:18-23).

peace and light, the south gave warmth, the west gave rain, and the north with its cold and mighty wind gave strength and endurance (1988:194)[33].

This passage also reminds us of an important sense in which the symbol of the medicine wheel cannot be reduced to simple dualities. For each of the four quarters is represented as a Power in its own right, with its own distinctive gifts, teachings and wisdom. When the two vertical directions or powers are included–the Sky above and the Earth below–there are a total of six such directions or powers. These are sometimes conceived of as forming a sphere, rather than a circle, as Frank Fools Crow points out when discussing the symbolism of the Sweat Lodge, which is a purification ritual common to many different First Nations. As he says of the dome shaped lodge, which is itself a representation of the structure of the world, "Its true shape is that of a ball, and the bottom half of the ball is underground" (1991:104)[34].

Thus, the relational symbolism of the circle, and the ideal of balance which it communicates, are far more complex than a simple duality is capable of evoking. Nor can an understanding of the teachings discussed above be reduced to oppositions alone, for this is to miss the particular characters and powers attached to each of the particular directions, and thus much of the meaning of the symbols. This point should be kept in mind throughout the following, particularly when the relational symbolism discussed in the

---

[33] A corresponding passage form Neihardt's original interview transcript reads: "It is the duty of the four quarters to nourish and strengthen the flowering tree at the center which represents the growth of the people in all its generations. The east gives the sunlight and peace, the south brings warmth, the west brings the rain to nourish, the north gives it a strong wind so that it may have strength and endurance, and all these so that the tree may grow and flower" (DeMallie 1985:240).

[34] For Lakota accounts of the inipi or purification ceremony see Black Elk (1971:31-43); Fools Crow (1991:103-14); Lame Deer (1972:164-71); McGaa (1990: 61-72).

present chapter is compared to the dominant philosophy of the Western world.

The last point to be made, before going on to compare this relational symbolism to that of the West, is to reemphasize that the circle is conceived of as a *natural archetype*. In other words, the model for human patterns of relationship is arrived at by looking to the natural world, and the way in which things in nature are related to one another. As Lame Deer observes, "We see in the world around us many symbols that teach us the meaning of life...To you symbols are just words, spoken or written in a book. To us they are part of nature, part of ourselves" (1972:96-97). Or as Bobby Woods once pointed out, while acting as spiritual counselor to a group of Native American inmates in a Canadian prison near my home:

> We're sitting in a circle today. Were replicating something that is natural, just like this universe. Just like the sun, the moon and the earth moving in harmony, never interfering with one another. A circle. And the people looked at that and said; 'We can't go wrong, if we used that natural way! And so they did.' (National Film Board of Canada 1990).

In attempting to understand life, and the proper way of living it, they do not abstract themselves from nature, but see themselves as an integral part of the world around them. For the medicine man, or practitioner of such a philosophy, "must be of the earth, somebody who reads nature as white men read a book" (Lame Deer 1972:160). To learn how to live correctly one must look to the world of nature, in which one is a participant, and learn from the patterns of relationship which can be observed among other living beings. In this way, as Ed McGaa relates, "We are learning directly from God's creation. There is no middle person to alter, or confuse, the direct perception of real, God-designed knowledge" (1990:17). The world is there for all to see, and one requires no special equipment to begin learning from the ways of all things; only nature itself, and time for contemplation, which Native American life styles traditionally provided in abundance.

From such a point of view, ethical problems concerning how one should live are not separated from more practical concerns such as finding food and shelter. As George points out concerning the views of his own Coast Salish people, "the woods always provided us with food, homes, comfort, and religion...Our church was nature" (1982:49 and 59). Or as Eastman expresses this point in his own poetic style:

> There were no temples or shrines among us save those of nature...the Indian...would deem it sacrilege to build a house for Him who may be met face to face in the mysterious shadowy aisles of the primeval forest, or on the sunlit bosom of virgin prairies, upon dizzy spires and pinnacles of naked rock, and yonder in the jeweled vault of the night sky...He needs no lesser cathedral (1980:5-6)!

Clearly, Native American philosophies seem to have made no sharp separation between the sacred and the secular, for as Cajete suggests, "[i]n the Native mind, spirit and matter were not separate; they were one and the same...They experienced nature as a part of themselves and themselves as part of nature" (2000:186)[35]. Their religion–like the food and water which sustained them, and the clothing which they wore–was seen as arising from nature itself. Religious precepts were not handed down from on high and preserved in sacred books, but arose from their daily experience of life in the natural world. And since nature itself was considered to be sacred, there was no need to look beyond it or transcend it in order to acquire knowledge of sacred or spiritual matters. As Fools Crow reiterates this point, "in the old days my people did not separate daily life in the world from spiritual life. Everything was spiritual. We were soaked with it. It is only now that we see a

---
[35] See also: Cajete (2000:218).

difference" (1991:50).

Due to the fact that the sacred is seen as immanent within the natural world, one can do no better than to look to nature itself when seeking successful patterns of life. For as Eastman describes the teachings he received as a child from his grandmother and uncle, while still living the nomadic, hunting life of his ancestors:

> After arriving at a reverent sense of the pervading presence of the Spirit and Giver of Life, and a deep consciousness of the brotherhood of man, the first thing for me to accomplish was to adapt myself perfectly to natural things–in other words, to harmonize myself with nature (1977:1-2).

Or as Black Elk related several decades earlier, "nothing can live well except in a manner suited to the way the Power of the World lives and moves to do its work" (1988:212)[36]. Both one's own behavior, and the patterns of relationship in the social sphere more generally, are to be adapted to those which can be observed in nature among other forms of life, which Cajete describes as "[m]imicking the processes observed in nature" (2000:101). Or as Deloria elaborated upon this theme, again more recently:

> The task of the tribal religion...is to determine the proper relationship that the people of the tribe must have with other living things and to develop the self-discipline within the tribal community so that man acts harmoniously with other creatures" (1994:88).

After all, society is here conceived of as a *part of* nature, and it must move according to the same patterns as those in nature if it is to be successful. The similarity of this type of thinking with the

---

[36] This is Neihardt's wording, but in light of Eastman's earlier writings he does not seem to be misrepresenting Lakota philosophy by giving Black Elk's vision a more explicitly ecological slant. Consider also the following, which is Black Elk's recollection of the words of Red Dog, an older holy man who helped him to enact part of his great vision on Earth in about 1884: "Boy, you had a great vision, and I know that it is your place to see that the people might walk the road in a manner satisfactory to all its powers. It is the duty for you to see that the people will lead and walk the right road, because if it is not done, in the future our relatives-like will disappear" (DeMallie 1985:240).

ecological concept of adaptation to the natural world could hardly be more striking. Indeed, beginning several decades after the publication of Eastman's and Black Elks words, adaptation has come to be seen as the fundamental behavioral norm among eco-holists as well, as shall be discussed in more detail in Sections III and IV.

3. THE SKY KING: DUALITY AS SEPARATION

In the older mother myths and rites the light and darker aspects of the mixed thing that is life had been honoured equally and together, whereas, in the later, male-oriented, patriarchal myths, all that is good and noble was attributed to new, heroic master gods, leaving to the native nature powers the character only of darkness–to which, also, a negative moral judgment now was added. For, as a great body of evidence shows, the social as well as mythic orders of the two contrasting ways of life were opposed (Campbell 1964:21).

JOSEPH CAMPBELL argues that the mythology of the Goddess was dominant throughout the Paleolithic period, even in Europe. And as the previous sections suggested, similar types of philosophies continue to be espoused by Native Americans and other indigenous peoples up to the present day. Campbell sees a new and contrary mythology beginning to emerge in the Neolithic, and flowering in the context of the early civilizations of the Middle East, to which he traces the origins of Occidental mythology and religion. Previously myths had emphasized divinity as immanent within nature, the symbolism of the circle, and of the cycles of life and death, light and dark, which unite the opposites. This begins to be replaced by mythic themes emphasizing a transcendent male God, the separation of the opposites into opposing principles, and the denigration of that side which had been most closely associated with the Goddess, the Earth and the Female.

While in the villages of the early Neolithic, Goddess myths and symbols continued to thrive, as centralized states developed these forms of religious expression were "radically transformed, reinterpreted, and in large measure suppressed, by those intrusive patriarchal tribesmen whose traditions have come down to us chiefly in the Old and New Testaments and in the myths of Greece" (Campbell 1964:7)[37]. This earlier mythology originally began to

---

[37] Campbell estimates the origins of Occidental mythology at circa 1250 B. C.

be suppressed and replaced in various contexts in the Middle East, with Judaism, Christianity, Islam and Mormonism being surviving variants and, along with ancient Greek mythology and philosophy, the sources from which their secularized heirs of the present day eventually arose[38].

Eliade traces the associated change in Occidental relational symbolism from cyclical to linear conceptions of time, to essentially the same context from which Campbell derives the origins of Occidental mythology. Earlier myths had always represented time as cyclical, which is common in Native American thought as well. Modeling their symbolism after such natural temporal cycles as that from light to dark to light in the passing of the days; that of the waxing and waning of the moon to mark out months; and that of the great cycle of the seasons in the passing of years; longer periods of time were also often represented as cyclical. Often, the past was thought of as a series of cycles or ages leading up to the present age, which would itself come to an end, and in its passing give birth to the next.

This cyclical symbolism–which Eliade designates "the myth of the eternal return"–begins to be replaced in the Levant among the ancient Hebrews by a more linear, non-repetitive, historical conception of time (Eliade 1954)[39]. As the Bible tells us, by creating the

---

[38] While aspects of a new relational symbolism appear in early Occidental religions, variants of a more Goddess oriented, animistic view continued to exist in Europe up until the advent and of modern scientific philosophy. Carolyn Merchant discusses a variety of premodern European beliefs which continued to view the Earth as feminine and alive, though such views were not necessarily condoned by the dominant Catholic dogma of the time (1980:1-41).

[39] The contrast between linear and cyclical conceptions of time does not imply an absolute dichotomy between the two, but is a matter of relative emphasis. People with a more cyclical conception of time can distinguish between before and after, while those with a more linear conception still recognize the cyclicity of the seasons. Lame Deer (1972) offers the metaphor of time as a spiral, a metaphor which balances both the linear and cyclical aspects of time.

Earth in six days, and resting on the seventh, God himself provided the archetype for the seven day week, the basis of our present calendar. And this history of the people–which now describes specific historical events and persons, and places them within a linear temporal framework–is no longer maintained and passed down by an oral tradition. Instead, it is written down in the Sacred Scriptures, where it is directly accessible only to a hierarchically organized, literate priesthood, the only ones who can read the Word of God and interpret it for the illiterate majority. As Campbell relates:

> The high function of Occidental myth and ritual...is to establish a means of relationship of God to Man and Man to God. Such means is furnished...by institutions, the rules of which cannot be learned from any scrutiny of nature...these have to come from God himself, as the myth of each institution tells; and they are administered by his clergy (1964:4).

The common people no longer have any direct access to knowledge of God or His laws, and they can no longer look to nature and the ways of all things in order to understand the sacred on their own, for to do so would be to practice idolatry. God begins to be seen as *beyond* nature, as an exclusively transcendent deity who intervenes in events, when He does, from the *outside*–from *above*–while the theme of *immanence* is played down. As Eliade reemphasizes, this new view is "the exclusive creation of a religious elite" (1954:107). This elite group now stand in a position of social power, where only the members of the priestly hierarchy have direct access to the Word of God, and act as interpreters of the Divine Will for their congregations. In a largely illiterate world, even their ability to *read* His word must have been rather awe inspiring in the eyes of the common people, and the priests have now become the indispensable intermediaries in the relationship between God and His flock.

This pattern was reproduced in Europe, as well, where it was the norm throughout the Middle Ages after the diffusion of Christianity into Rome, with the eventual collapse of that empire leaving the Holy Roman Church in its place. After remaining dominant for centuries, the Pope's authority as the exclusive Earthly interpreter of the Divine Will was finally challenged by three related, and roughly contemporaneous developments: the emergence of Protestantism; the advent of printing and a consequent growth of literacy; and the early stirrings of the Renaissance. Yet the dominant Western world view at present–that of modern science–which is the eventual result of this shift of values, and the new ways of life consistent with them, retains a strikingly similar relational symbolism, as shall be illustrated below.

Also closely tied to the idea that divinity is transcendent, or located beyond the observable world of nature, is the picturing of nature–now rendered profane–in a negative and inferior light. For it is no longer nature to which one turns for a model of the sacred, but to the structure of social institutions–the emerging hierarchy.

A mythic model of this entire transformation is found in the legend of Tiamat and Marduk. This story is a legend of early Mesopotamia, which is also the first place in which writing developed, which dates to the reign of the emperor Hammurabi (circa 1728-1686 B. C.). In this early example of Occidental mythology, the celestial symbol is no longer the moon, with its cycle from dark to light embodying the unity of opposites, "but the golden sun, the blaze of which is eternal and before which shadows, demons, enemies, and ambiguities take flight." Attention shifts "to the foreground figures of duality and combat, power, profit and loss, where the mind of the man of action normally dwells," thus being a very masculine metaphor. Marduk is appointed king of the gods of heaven after being chosen as their champion in a forthcoming war against the Earth Goddess–Tiamat–and her allies. In fact, he demands the position for undertaking this task,. And thus "the art of politics, the art of gaining power over men, received for all time its celestial model." The connection between the origins of the transcendent Sky God metaphor, and the attempt to justify both social and patriarchal hierarchies is also made more explicit.

When Marduk faces Tiamat in battle she is quickly vanquished by his superior might, and he "split her, like a shellfish, in half," setting one half above to form the Sky, and one below to form the Earth[40]. Yet while the Goddess thus remains immanent within the creation, both She, and it, are soon to be denigrated. Indeed, the Earth Mother has already been slain as an enemy of the transcendent sky gods.

After his victory, Marduk then proceeds to divide the gods into the Good and the Evil. The latter–not surprisingly–are Tiamat's allies. It is also from these latter that he fashions human-kind, who "will be required to serve the gods." For the gods are now conceived of after the model of Kings, who rule the world from a transcendent heaven. The emperor Hammurabi himself is then appointed to his position by the triumphant sky gods, thus to rule by divine right from the highest Earthly position in a hierarchy which has its apex in heaven (Campbell 1964:73-85).

Anyone familiar with European history should immediately recognize the similarity to the Medieval doctrine of "the divine right of kings," which long enjoyed the official sanction of the Church of Rome. Clearly, it is no coincidence that Christians have long referred to both their heavenly ruler and their Earthly ones as

---

[40] Thus, Marduk's destruction also results in the creation.

"lord," for their understandings of the former have generally been modeled after the latter. As Campbell concludes:

> In the older view the goddess Universal was alive, herself organically the earth...Now she is dead, and the universe is not an organism, but a building, with the gods at rest in luxury: not as personifications of the energies in their manners of operation, but as luxury tenants requiring service. And Man, accordingly...as a robot fashioned to serve (1964:82).

Nature is not only seen as profane, or even evil, but as *dead*. It becomes a dangerous wilderness which must be fought and conquered rather than the fertile, life-giving realm of the Goddess, and "all these wild, uncultivated regions and the like are assimilated to chaos" (Eliade 1954:9). It is only that which has been ordered by humanity which is considered ordered, while those orderly patterns which spontaneously arise from nature, or from the Goddess, are denigrated and dismissed.

This denigration of nature, which Cajete describes as "biophobia, our basic fear of nature" (2000:153), and the accompanying elevation of the social, is also reflected in a sort of repositioning of the center of the world. In previous mythology, the *axis mundi*–the sacred center which provided the bridge between Heaven and Earth–was often represented by such natural phenomena as a sacred mountain, or by the World Tree which stood at the center of the Earth. As we have already seen, the symbolism of the World Tree growing at the center of the universe was also common in Native American religions and rituals, particularly in the Sun Dance, and in Black Elk's great vision. Indeed, even in the Biblical story of Eden the World Tree stands at the center of the garden. Here its meaning is transformed, however, since God himself forbids Adam and Eve to eat of its fruit, and to partake of the knowledge which it provides. When they do eat of it despite His injunctions He then casts them out of the Garden, where they had originally lived a happy and carefree life, and they and their descendants immediately set about founding intensive agriculture and the first centralized states.

In this myth the symbol of the World Tree came to be seen as the source of the knowledge of Good and Evil, and eating of its fruit as the Original Sin of humankind. The sacred center–the link between Heaven and Earth–now came to be assimilated to the temple complex and its resident priests. And surrounding the temple, of course, was that model of humanly constructed order–the city–the political and economic center of the state. So the "centre" came to be equated with the seat of social power, and it is from this centre of religious and political authority that order flows, an order which must be imposed by "society," or the elite, upon an unruly world, rather than arising spontaneously from nature. Indeed, because pristine nature has come to be equated with chaos, "[s]ettlement in a new, unknown, uncultivated country is equivalent to an act of Creation" (Eliade 1954:10); bringing about, as it does, a country ordered by the hand of Man in place of a natural but fearful wilderness. This is the precise opposite of the previous views, as it is of those of various Native American peoples. As Standing Bear observes, for example:

> We did not think of the great open plains, the beautiful rolling hills, and winding streams with tangled growth as 'wild.' Only to the white man was nature a 'wilderness' and only to him was the land 'infested' with 'wild' animals and 'savage' people. To us it was tame. Earth was bountiful and we were surrounded with the blessings of the Great Mystery. Not until the hairy man from the east came and with brutal frenzy heaped injustices upon us and the families we loved was it 'wild' for us. When the very animals of the forest began to flee from his approach, then it was that for us the 'Wild West' began (1978:38).

From the various contrasts considered above–such as the opposition between culture and nature, or between Heaven and

Earth–it would seem that the rise of Occidental mythology was accompanied by a new way of conceiving of dualisms, and a new way of representing them symbolically. As Campbell suggests:

> The patriarchal point of view is distinguished from the earlier archaic view by its setting apart of all pairs-of-opposites–male and female, life and death, true and false, good and evil–as though they were absolutes in themselves and not merely aspects of the larger entity of life (1964:26-7).

Where previously they had been considered as complementary aspects of a larger unity, it is as though a line has now divided the cycle which once united them into two separate pieces, which now exist as self-sufficient entities. They are now conceived of not only as basically separate, but as fundamentally *opposed* to one another. In this way male has been separated from female, "the subject and the object are indeed two, separate and not the same–as A is not B, as death is not life, virtue is not vice, and the slayer is not the thing slain" (Campbell 1964:78).

Where previously the ethical ideal had been represented as the point of balance between the two extremes, one of the extremes is now singled out to serve as the ideal–to be maximized and given social approval–while the other is to be minimized, and is socially scorned. The Judeo-Christian concept of Good and Evil as two absolutes in perpetual conflict with one another remains one of the best examples of this symbolism. Any actions not sanctioned by the Church and the Scriptures are considered, always and everywhere, to be Sin. Here the profane is separate and opposed to the sacred; and where in the earlier mythology–as well as in the views of many Native Americans up to the present day—the Sky and the Earth had been honoured as complementary aspects of the Great Mystery, now God and the absolute good are associated only with the transcendent Heavens, while below lies Hell and the realm of Evil. So the direction which had previously been associated with the immanent Goddess has come to be associated with evil, and has even come to be represented by a male demon–Satan, Lucifer, the Devil–the fallen angel who is seen as opposed to the male God of heaven. In this way the divine principle is removed from the creation, and the Earth itself is rendered profane, while Woman is largely removed from religious symbolism[41].

Similarly, the *mind*, which is generally equated with the concept of the transcendent soul, is separated and opposed to the body, and the bodily appetites and desires are scorned. Sex, for example–the joyous coming together of the opposites in an act of creativity–was considered little better than a necessary evil by the Church of Rome throughout the Middle Ages, or as an act which only the holy sanction of the church could render acceptable, and was considered a sin out of holy matrimony[42]. The body must also remain covered and hidden from sight, as seeing it might incite the Passions. For when they ate of the forbidden fruit from the tree of knowledge, one must remember, Adam and Eve also became ashamed of their nudity–*or perhaps their humanity*.

This denigration of the bodily is still embodied in the swearing of the English language, or in those words which are not to be uttered in "polite" company, most of which refer to sex and other bodily processes. Unlike the West, few First Nations peoples seem to have had such a category of profane concepts, which were to be carefully separated from "acceptable" language. For as Joseph Dion states concerning the Cree language, "[a]bout the only thing we haven't got and which we have to borrow from other tongues, is real honest-to-goodness swear words" (1979:2). Similarly, as Lame Deer relates, in the Lakota language:

---

[41] She does, however, make a covert return in the Christian faith in the form of Mary, the Mother of God.

[42] As it is by some Christian churches in the present day.

We can't curse. We have no four-letter words. Sure, we have a word for intercourse; it means just that–a man and a woman making love. That's hard to understand for us–using a word that really means bodies coming together in joy–using that for a curse (1972:204).

Yet throughout Western history this union of the opposites in an act of creation has generally been rendered in a negative light, as has Woman. Even now, priests in the Catholic Church continue to be required to maintain their celibacy, so that they may remain untainted by such profane activities, and devote their lives to the "higher" spiritual and intellectual pursuits. And since Woman has generally been conceived of as most closely associated with sex and the body, since it is She who excites the passions of Man, who gives birth, and who suckles the young, so also Her way of being was also considered to be inferior to that of men.

Indeed, among the upper classes of Europe it was long the practice to employ wet nurses to release noble women from so profane a task as breast feeding their own young, though this has now been widely replaced by bottle feeding. Similarly, a woman's first menstruation is no longer represented as a sacred event which shows that she has acquired a creative power analogous to that of Mother Earth, for following traditional Christian symbolism, it is now often called "the curse," and is seen as a lasting reminder of God's punishment of Eve.

A secularized example of this relational symbolism, which is more consistent with the dominant attitudes of the present day, is to be found in the utilitarian ethics of Jeremy Bentham and John Stuart Mill. Here it is *pleasure* which is always and everywhere good, while *pain* has taken the place of absolute evil. In this *hedonistic* view it is pleasure which is to be maximized, while pain is to be minimized, and by quantifying the total amounts of pleasure and pain which will result from an action, the one can be subtracted from the other, with the result of this equation being the correct course of action–that which results in the greatest amount of general pleasure.

This is again to be contrasted with various First Nations traditions, where pain as well as pleasure are often acknowledged as complementary aspects of life. Pain is even seen as useful for some purposes, and sometimes plays an integral role in religious ritual. As we have already seen, in the self-sacrifice of the Sun Dance the participants offer their own flesh and pain so that the people and the entire creation might continue to live and flourish, and so that they might gain in understanding through the gift of sacred visions. Abstaining from food and drink for several days, and sometimes flesh offerings by relatives on one's behalf, are often important parts of the quest for vision as well. As Igjugarjuk an Inuit shaman points out, in the view of the shaman:

> True wisdom is only to be found far away from people, out in the great solitude, and it is not found in play but only through suffering. Solitude and suffering open the human mind, and therefore a shaman must seek his wisdom there (as cited in Halifax 1979:69).

The theme of separation in Western philosophy is not only represented by the separation and opposition of dualities, however, but finds expression in the related doctrine of atomism as well. The earliest formal articulation of this atomistic view–that the parts are somehow *prior* to, or more *basic* than, the relationships which exist between them–can be traced to the works of Democritus, a pre-Socratic philosopher of ancient Greece. A modern version occurs in the works of René Descartes, who is widely considered to be the founder of contemporary modern philosophy, and who was one of the first to articulate the scientific world view and its distinctive methodology. In his *Discourse On Method* this atomistic assumption appears as one of the four steps in his proposed method of "rightly conducting one's reason and seeking truth in the sciences," where he resolved "to divide each of the difficulties I was examining into as many parts as possible and as is required to solve them best" (1980:10).

From such a point of view, reality is no longer considered to be a basic unity where, while differentiation of parts is acknowledged, the larger patterns of relationship which unite them into a larger whole are also considered. Instead, it is only the parts which are considered as having any fundamental reality, while larger wholes are considered to be mere epiphenomena–as nothing more than the sum of their parts. As David Hume once expressed this atomistic tendency in modern Western thought, "there appears not, throughout all nature, any one instance of connection which is conceivable by us. All events seem entirely loose and separate. One event follows another, but we never can observe any tie between them" (1962:87).

Thus, the separation of a consideration of the parts from that of the whole, and the granting of priority to the parts, reinforces the theme of separation in a further respect. And this is still the manner in which modern science proceeds: breaking phenomena down into parts, separating, classifying and quantifying; measuring length, breadth, height, temperature, time, motion, volume, energy and weight, so as to see how the parts may be manipulated and reassembled to suit the experimenters ends. The utility of such a procedure from the perspective of mechanical engineering and applied science should also be apparent.

Neither was the Western habit of measuring and subdividing everything it encountered lost upon Native American observers. Eastman, for example, recalls his uncle speaking of the white man's habits and commenting upon the fact that "[t]hey have divided the day into hours, like the moons of the year. In fact, they measure everything. Not one of them would let so much as a turnip go from his field unless he got full value for it" (1971:242). This was not the way of Native American peoples. Instead, as Carl Sweezy points out with reference to Western measurement of time:

> Hours, minutes, and seconds were such small divisions of time that we never thought of them. When the sun rose, when it was high in the sky, and when it set were all the divisions of the day that we had ever found necessary when we followed the old Arapaho road (as cited in Nabokov 1991:210).

The day was not seen as divided into pieces, advancing in rhythm with the mechanical ticking of the clock, as Hume appears to have conceived of it. Rather, time was divided only by the natural cycles of celestial motion. The usefulness of such artificial, socially standardized subdivisions in a culture where, as people say, "time is money," is again evident, however. For of course, work time must be carefully separated from leisure time, so that the former may be quantified and converted into dollars. Yet from Native American perspectives such absolute separations of work and play, public and private, or sacred and profane seem to have been entirely foreign, for they simply had no use for them. Perhaps the words of Leonard Crow Dog provide an appropriate summary of such points of view:

> Life is holiness and everyday humdrum, sadness and laughter, the mind and the belly all mixed together. The Great Spirit doesn't want us to sort them all out neatly. He lets the white people do it (as cited in Halifax 1979:77).

4. SOCIAL ARCHETYPES: THE PYRAMID AS A RELATIONAL SYMBOL

since nature makes nothing imperfect or in vain, she must have made all other things for the sake of men. For this reason, the art of war, too, would be by nature an art of acquisition in some sense. For the art [of acquisition] includes the art of hunting which should be used against brutes and those men who, born by nature to be ruled, refuse to do so, inasmuch as this kind of war is by nature just (Aristotle 1986:557).

IN DOMINANT WESTERN ways of thinking, dualisms are not only conceived of as being fundamentally separate and opposed to one another. Choosing one of the extremes as the ideal carries with it the implication that this side of the duality is *superior* to its contrary as well. When applied to social relationships, this way of thinking sets up a rudimentary hierarchy whereby the superior or "higher" type is justified in controlling its inferiors. Karen Warren notes a clear link between such value dualisms and the larger hierarchies which flow from them in her account of what she terms "the logic of domination" (1990:125-46)[43]. By adding more "higher" or "lower" levels, the simple dualism which places ruler over ruled, for example, quickly yields the pyramidal form of a social hierarchy.

In speaking of social hierarchies (or hierarchies of *power*) in what follows, I would like to clearly distinguish them from *classificatory* hierarchies, such a those used in biology to classify the various types of living things into species, genus, family and the like. For while both are represented symbolically in the same manner, their common pyramidal form obscures a fundamental difference. The relationship between genus and species, for example, is that of a smaller category located within a larger, which

---

[43] Her idea of the logic of domination is quite similar to Murray Bookchin's earlier consideration of "epistemologies of rule," which he developed in *The Ecology of Freedom* (1991).

is hardly congruent with that between ruler and ruled, which describes a difference in *status* and *social power*. In a *social* hierarchy the "upward" direction is from *inferior* towards *superior*. "Superior" carries the further implication of the greater "value" or "worth" of these "higher" types, and the "higher" is placed in a position of "authority" over the "lower" as a result.

In the biological classification of species, however, no such implications follow. For in the latter case the "upward" direction is from *less inclusive* to *more inclusive* and, if the complexity of the system permitted, it could be better represented by returning to the symbolism of "circles within circles" suggested by Lame Deer, since each "higher" category *subsumes* those below it. Thus, where the "upward" direction in a *social* hierarchy is progressively more *exclusive–separating* types by virtue of differences in power, authority and value–the same direction in a classificatory hierarchy is more *inclusive–bringing together* various types by virtue of similarity, and illustrating their relationships to one another within a larger pattern of relationship. In the simplest terms, the difference between social and classificatory hierarchies (or *holarchies*), may be rendered as a contrast between a map of *differences in social power*, and a model of the relationships between *more inclusive, and less inclusive categories*, or patterns of relationship, respectively. For if we were to confuse the two relational symbols, we might end up equating a ruler with an entire society, or claiming that a genus has greater status and authority than a species!

There can be little doubt that social hierarchy has been the typical relational pattern of Western cultures as well as ancient states, however. Such a pyramidal structure of command and obedience has remained the exemplary model for social relations throughout Western history–from the hierarchy of the Church of

Rome and the kings to which it gave divine sanction to rule in the middle ages, to the bureaucracies of centralized governments, transnational corporations and military regiments in the present day. Indeed, such social hierarchies are even one of the defining features of a "civilization" (or state) in the anthropological literature.

Examples of the type of thinking which Warren describes are easily found throughout the dominant Western philosophy. As she points out, the key assumption in their arguments is the following: "For any X and Y, if X is morally superior to Y, then X is morally justified in subordinating Y" (1990:129). In such arguments, the duality which has often served as a starting point historically has been that between mind and body, or the intellect and the passions. The former in each pair is considered superior, while the latter is often being considered the road to evil. As Plato asks, for example, using metaphors highly reminiscent of Christianity:

> what is the origin and purpose of the conventional notions of fair and foul? Does not the one subject the beast in us to our human, or perhaps I should say our divine, element, while the other enslaves our human nature to the Beast (1974:417)?

From this passage it is quite evident that the "divine element" to which he refers is the intellect, while "the Beast" can only represent the passions of the body. The equation of the intellect with the "fair" and the passions with the "foul" is also clear. Aristotle, Plato's contemporary, would not disagree on this point. As he states, "an animal consists first of soul and body, of which the soul by nature rules but the body [by nature] is ruled...among men who are vicious and disposed to vice it is the body that often rules the soul" (1986:552). Unlike many Christian or modern thinkers, however, Aristotle attributes a soul to animals as well as humans, with this "irrational soul" being its life principle, which differentiates a living body from a corpse. Humans, however, are also seen as having a "rational soul," which is ranked higher than that which it shares with the other animals, just as the theoretical activities of the rational soul are ranked above its practical activities.

In the Middle Ages one finds Saint Augustine building a very similar hierarchy, which ascends from the inanimate, to the animate, to the intellectual, to God. Just as "that which has life is superior to what merely exists," so also that which thinks is superior to that which only lives, as God is superior to all. And since "no one doubts that he who judges is superior to that over which he exercises judgment," then God, who judges over people, and the mind, which judges over the passions, are each ranked above their inferiors (1953:142-43). One must remember, of course, that Augustine was writing during the Middle Ages, when there were few indeed who doubted the authority of their rulers, since they had no experience or knowledge of any other way of life. The literate classes themselves were largely members of the upper class as well, which makes their adherence to this model less surprising, for it is a justification of the superiority which they claim for *themselves*. As with Plato and Aristotle, Augustine also considers the influence of the passions to lead to evil, for "all evil deeds are evil for no other reason than that they are committed from lust" (1953:117). So as Aristotle summarizes the argument thus far:

> For those who adopt our division of the soul it is not unclear in which of those parts the end of man ought to be said to lie. For that which is inferior exists always for the sake of the which is superior, and this is similarly evident in things according to art and things according to nature; and the part which has reason is better (1986:600).

It remains, despite the evident similarity with social hierarchies, to apply this logic to social relations. This is generally done by arguing that one's social inferiors are less *reasonable* than oneself, and just as reason is considered to be superior to the passions, so also the more reasonable person is superior to the less reasonable. Thus the argument comes full circle, with the hierarchy of mind over body being used to justify the very social hierarchy on whose model it was originally constructed.

For example, Aristotle offers the following by way of justification for the three types of rule he observes in the Greek household–those of master over slave, husband over wife, and of a father over his children: "The slave does not have the deliberative part of the soul at all; the woman has it, but it has no authority; and the child has it, but it is not fully developed" (1986:564). Thus the slave, like other animals, is lacking the "higher" part of the soul, and therefore "it is by nature that some men are slaves but others are freemen, and that it is just *and to the benefit* of the former to serve the latter (1986:553, emphasis added).

Not surprisingly, it is the uncivilized "barbarians"–the non-Greeks–who are "slaves by nature," as are all people who refuse to be ruled over in a hierarchical fashion like properly "civilized" persons. One notes that the same logic which he uses to justify slavery is also used to justify patriarchy (itself a specific form of hierarchy), with the male conceived of as ruling over his wife and children as though the family itself were a little kingdom.

It appears that the Bible would agree, for as a punishment for Eve's original sin, God himself once stated that "in pain you shall bring forth children, yet your desire shall be for your husband, and *he shall rule over you*" (Genesis 3:16, emphasis added). Or as Aristotle puts it, "the male is by nature superior to the female, and [it is better for] the male to rule and the female to be ruled," since "the rule by both parts alike or by the inferior part is harmful to all" (1986:552).

So by conceiving of the intellect as superior to the body, and by arguing that their own intellect is superior to that of women, foreigners and children, they convince themselves of their own superiority (which none of them seem to have doubted), and of their own right to rule. As Aristotle saw it, "To rule and be ruled are not only necessary, but also beneficial" (1986:550). What remains unspoken, however, is the fact that it is primarily beneficial to the ruler. For as Plato relates; "it is better for every creature to be under the control of divine wisdom. That wisdom and control should, if possible, come from within; failing that it must be imposed from without" (1974:418).

Where Native American philosophies tend to turn to nature for inspiration, therefore, and seek to model social patterns of relationship after natural archetypes, dominant Western modes of thought appear to do precisely the reverse. For if other humans may be considered one's inferiors due to their presumed lack or inferiority of reason, how much more so must plants, animals and the rest of nature, which are incapable of speech, and consequently cannot attain to reasonableness at all? From the opposition which places mind over body, then, that which places culture over nature invariably follows, since human cultures are the only relationships among beings gifted with a soul or reason, rendering them superior to all other forms of life. Are we not told in the Bible that Man alone is created in God's image? And did not God himself command Adam to "[b]e fruitful and multiply, and fill the earth *and subdue it*, and have *dominion* over the fishes of the sea and over the birds of the air and over every living thing that moves upon the earth" (Genesis 1:28, emphasis added)[44]? Thus, as Thomas Aquinas once stated, in dominant Christian interpretations, "nothing subsisting is greater than the rational soul, except God" (1945:176).

Such a philosophy is unlikely to value natural archetypes, for this would be to model ones behavior after one's *inferiors*. Consequently, we find that in Western thought the natural world tends to be afforded no ethical regard whatsoever. Instead, it is seen as existing merely for the use of humanity. As Leading Cloud so perceptively observed, "White people see man as nature's master and conqueror, but Indians, who are close to nature, know better" (as cited in Erdoes and Ortiz 1984:5). Similarly, Standing Bear describes this contrast in the following way:

> The Indian and the white man sense things differently because the white man has put distance between himself and nature; and assuming a lofty place in the...order of things has lost for him both reverence and understanding...And here I find the great distinction between the faith of the Indian and the white man. Indian faith sought the harmony of man with his surroundings; the other sought the dominance of surroundings...For one man the world was full of beauty; for the other it was a place of sin and ugliness to be endured until he went to another world...But the old Lakota was wise. He knew that a man's heart, away from nature, becomes hard; he knew that lack of respect for growing, living things soon led to lack of respect for humans too. So he kept his youth close to its softening influence (1978:196-7).

In Western cultures, on the other hand, it is the hierarchical

---

[44] In addition to serving as a justification for subduing nature, this passage sometimes served as a justification for stealing Indian lands as well. For example, precisely because Native Americans did not "subdue" nature, the Massachusetts Court once stated that they did not own their land, for "only what landes any of the Indians, within this jurisdiction, have by possession or improvement, by subduing of the same, they have a just right thereunto, according to that Gen: I:28." See Cronon (1983: 63). In other words, only those few lands which First Nations cultivated in corn or other crops were seen as properly belonging to them, since "subduing" nature–as commanded by God–was seen as a prerequisite to owning it. Among the earliest critiques of Christianity's doctrine of dominating nature in the eco-holist literature is the work of Lynn White, Jr. (1968), which is cited with approval by Deloria (1994:83) in support of his own critique of the Western domination ethos.

structure of the social sphere–its command and obedience, and its superiority and inferiority–which is extended to its relationships with the natural world. As Augustine once articulated this view, "there is no order at all, where the better is subordinated to the worse" (1953:123). The concept of "order" itself is equated with social hierarchies, and the task is not to look to nature and adapt oneself to it patterns of interrelationship, but to *rank* its components according to their relative value from a human perspective, just as a social hierarchy ranks the members of society. Humanity is seen as superior to animals, which are superior to plants, which are superior to the "dead" soil, water and sunlight from which they grow. And just as peasants and serfs must render service to kings, so also each "lower" type in nature exists only to be exploited by the "higher." So as Murray Bookchin observes, "The notion that man must dominate nature emerges directly from the domination of man by man" (1986:95), for they follow precisely the same hierarchical logic. Such a sentiment is quite evident in the writings of Descartes, for instance, as when he states that through applying his scientific methodology:

> it is possible to arrive at knowledge that is very useful in life and that in the place of the speculative philosophy taught in the schools, one can find a practical one, by which, knowing the force and the actions of fire, water, air, stars, the heavens, and all the other bodies that surround us, *just as we understand the various skills of our craftsmen*, we could, *in the same way, use these objects* for all the purposes to which they are appropriate, and thus make ourselves, as it were, *masters and possessors of nature* (1980:33, emphasis added).

One feels like adding: "Nature also?" For the example which he offers as a model for the "mastery" of nature is unmistakably the preexisting mastery of one person or class by another in the social hierarchies of his time. According to pattern, Descartes also arrives at this position by positing an absolute "separation" between mind and body; labeling them "thinking substance" and "extended substance." Yet where the Greeks and the Medievals had at least attributed an irrational soul or life principle to animals, which raised them above "inanimate" rocks and rivers, their life has now dropped out entirely. It is no longer a life principle which animates our own bodies either, but the *mind* or *reason*, which no other animal shares. In Descartes' view our essential selves have come to be equated only with this transcendent mind, which is conceived of as *above* or *beyond* nature. In Aristotle's' account the mind had, at least, remained firmly a part of nature, even while seen as the superior part. Yet for Descartes superior or "above" came to imply transcendence as well. Similarly, as Peter Miller suggests that:

> The contours of Western humanism have become sharply defined...since the time of Descartes. The multi-layered hierarchies of value in the classical and medieval worlds have been leveled to just two elevations: a plain below and a plateau above...the human circle on the plateau...while, on the plain below, the rest of the universe is spread: animal, vegetable, and mineral yoked equally in the service of humankind (1983:319-20).

So while the earlier hierarchical symbolism had always been consistent with the value dualism which placed mind over body, this symbolism was not completely reductive. As with the symbolism of the six directions in Native American philosophies[45], the hierarchy of various "levels of being" could not be reduced to simple dualisms. But in the picture Descartes paints *all* differences in value are reduced to only two kinds, and everything beyond the human sphere exists merely to be manipulated and used for which ever ends and purposes its owners may choose. Thus, his extreme form of reductive dualism also obscures the connection between such value dualisms and the social hierarchies with which they are consistent. Yet whether reduced to a simple duality or a more complex hierarchy, this type of symbolism has always tended to have

[45] The Four Winds, plus Earth and Sky, as discussed above.

the same implications, and the whole picture appears to be contrary to Native American views. For as Eastman relates:

> The first American mingled with his pride a singular humility. Spiritual arrogance was foreign to his nature and teaching. He never claimed that the power of articulate speech was proof of superiority over the dumb creation; on the other hand, it is a perilous gift (1980:88-9).

Or as Cajete has expressed this point more recently, in Native philosophies, "[k]nowledge must be both a source of joy as well as one of gravity and respect, because responsibility to the life that surrounds us is ignored only at great peril" (2000:104). Unlike Western points of view, where difference has generally been taken to imply superiority and inferiority, Native views leave room for the possibility of conceiving of difference as complementarity. Since they are not under the sway of hierarchical models of relationship, difference need not imply an absolute separation into better and worse, nor that the superior is justified in exploiting the inferior.

So where Native American traditions tend to follow natural archetypes–looking to natural cycles and interrelationships for a model after which social relationships may be patterned–dominant Western philosophies have tended to follow social archetypes, with the model for *all* relationships being the hierarchical structure of command and obedience embodied in its own patterns of social organization. This pattern is then imposed upon the natural world as well, lying as it does at the base of this great pyramid. To put it most simply; where First Nations literatures often seem to propose that the aims and purposes of society should be adapted to nature, dominant Western philosophies tend to suggest that nature exists for no other reason than to be adapted to the aims and purposes of society. Though they did not express it in these terms, many Native Americans were not unaware of this difference even a century ago, as is evident from the words of Chief Flying Hawk, when contrasting the mobile lifestyle of his Lakota people with the permanent dwellings of the West:

> If the Great Spirit wanted men to stay in one place He would make the world stand still; but he made it always to change, so birds and animals can move and always have green grass and ripe berries, sunlight to work and play, and night to sleep; summer for flowers to bloom, and winter for them to sleep; always changing; everything for good; nothing for nothing. The white man does not obey the Great Spirit; that is why the Indians never could agree with him (as cited in McLuhan 1971:64)[46].

---

[46] Flying Hawk belonged to the Oglala clan and fought alongside Crazy Horse at the Battle of the Little Bighorn in 1876. He was born in 1852 and became a chief at the age of 31. This passage represents his opinions as an elder. He died at Pine Ridge in 1931. The passage was originally published in: McCreight (1947:61).

II

DOMINATION & ACCOMMODATION:
THE COLONIAL ENCOUNTER

1. THE INVENTED INDIAN?

> When two cultures meet, especially cultures as different as those of western Europe and indigenous North America, they inevitably interpret each other in terms of stereotypes...We have a long history of romanticizing Indians, discovering in their character and culture many fine qualities we think lacking in our own. From the Noble Savages of years ago to the Mystic Shamans and Original Environmentalists of today, we continue to create idealized images of Indians which may have as little connection to reality as the demonic ones (Francis 1992:221-22).

IN TURNING to a consideration of the colonial encounter, as it was enacted in the North American context, the present section shall not attempt to provide an exhaustive historical account. This is a task which has already been ably undertaken by many others[47]. Rather, the focus in what follows shall remain upon the relational symbolism of the two cultural types discussed above, and upon an abductive comparison of these relational symbols with the actual patterns of social and ecological relationship enacted within each, and in the encounter between them. Where the previous section focused upon symbolic activity in the ideological sphere, the focus shall now be expanded to consider the larger social and ecological contexts–to the manner in which these ideologies were enacted in practice historically, and to the forms of life from which they arose.

Of course, it was not simply a matter of "the West" meeting "the

---

[47] For a classic discussion of the colonization of the American west, see Brown (1970), and for a detailed consideration of the colonization of indigenous peoples more generally see Bodley (1999).

Natives." The cultures of Native America differed among themselves no less than did those of the various European nations which
settled their lands, just as the various individuals within each
nation differed among themselves. At the time of contact Native
patterns of subsistence varied widely: from an almost exclusive
reliance upon the hunting of game animals, whaling and fishing in
the Arctic, to mixed hunting, gathering and fishing, to horticulture
and intensive agriculture. Their political organization also ran the
gamut from small, loosely organized bands consisting of only a
few related families, to larger tribal groups and confederacies of
tribes such as those of the Huron, Iroquois and Powhatan, to farming village societies such as those of the Hopi and Tewa, to the centralized state of the Aztec Empire in Meso-America.

These differences notwithstanding, the majority of Native
American societies seem to have shared a fairly consistent world
view, as did the European societies. Yet each culture elaborated
upon these general patterns in its own way, so while the present
work takes a generalist perspective, in that it focuses upon the
larger similarities which characterized the Occidental and Native
American cultural types, the detailed ethnographic differences
provided by a particularist perspective should not be forgotten.

As both Berkhofer (1978) and Francis (1992) point out, popular
images of "Indians" have also varied considerably throughout the
history of the colonial encounter in North America. Francis
(1992:221-22) summarizes the history of these images by
suggesting that there were three stages to popular Western
conceptions of the Native peoples of North America. During the
early fur trade era, especially in New France, where large-scale
settlement was not pursued initially, images of the Indian were
largely positive, as shall be discussed below. This is not surprising,

given that carrying out the fur trade implied that the French not only made their living trading with indigenous peoples, but also that they relied upon them for protection and safe passage to conduct their trade.

When and where settlement increased, however, and when Native peoples began to grow antagonistic towards increasing encroachments upon their territories by settlers–and especially farmers–popular images of the Indian generally took a distinctly negative turn. This view attempted to justify their subjugation, and the taking of their lands. In the latter half of the past century, with the rise of ecological thought in the West, there has been a slow return to more positive images once again. This section shall examine two such positive images of Native Americans, each of which shall become important as the argument develops in Sections III and IV.

Previous works which have examined such popular images of "Indianess" in North America can be placed upon a continuum, based largely upon *whose* images of Indianess they discuss. At one extreme is the position of Francis, who largely confines himself to the image of Indians as understood by Western culture. Thus, he concludes his arguments with the statement that "The fantasies we told ourselves about the Indian are not really adequate to the task of understanding the reality of Native people" (1992:224).

Berkhofer, on the other hand, referring to the concept of "Indianess" itself, proposes that "What began as reality for the Europeans ended as image and stereotype for Whites, and what began as an image alien to Native Americans became a reality for them" (1978:195). More specifically, "the original White image of the Indian," has always been that of "a separate but single collect-ivity," and the relative power of the colonizing societies often forced

Natives "to be the Indians Whites said they were, regardless of their original social and cultural diversity" (1978:195). Thus, Berkhofer suggests that even the concept of "Indianess," was foreign to Native American societies, who identified themselves with their various tribal groups or Nations, not with what was essentially a racial category invented in the West. Yet the relative power of the dominant society has often forced Native peoples to adopt, in whole or in part, Western concepts of "Indianess" based on Western notions of Native Americans as a single, homogeneous group with a single homogeneous culture.

At the other extreme of the continuum is Clifton's work, which also discusses "the conventional story most commonly told about the Indians of the United States and Canada" (1990:23). Clifton describes this story as "a dominant narrative structure," or as "a preferred story-line about the Indian, and above all else about the relationships between the Indian and the Whiteman, past and present" (1990:19). In other words, what Clifton describes is a "politically correct" story about colonial history, and of the re-lationships between Native Americans and the settler population. Yet while he acknowledges the involvement of Western academics in promoting this image or story, Clifton suggests that it is an image promoted by indigenous peoples themselves, for:

> contemporary Indians are the primary political-economic royalty owners of this major work of cultural fiction...over the past several decades Indians have become central actors in editing and revising, in garnishing, enlarging, and serializing the narrative's substance, busily occupied with inventing their own preferred images (1990:19).

Unlike Francis and Berkhofer, Clifton focuses upon images of Indianess invented by indigenous peoples themselves, and suggests that much, if not all of this popular image of Indians and the colonial encounter was invented for their own political purposes. Clifton appears to argue that many of these images are directly attributable to the Pan-Indian movement of recent decades, which encourages Indians to identify themselves, first and foremost, as "Indians," and only secondarily as members of the various First Nations and tribal groups. Thus, Pan-Indianism emphasizes a generalist approach similar to the present work, in that it attempts to focus upon the common themes which crosscut Native philosophy, rather than upon the specific traditions of particular peoples.

And while, as Berkhofer points out, this does represent an adoption of an originally Western way of thinking of indigenous peoples, the Pan-Indian emphasis of much contemporary Native American scholarship could just as easily be characterized as a philosophical school, as it has been by McPherson and Rabb (1993:1-3), which is attempting to illustrate and articulate the common themes which characterize a particular type of philosophy. We commonly speak of "Christian philosophy"–despite the many sects and denominations of that faith, for example. Yet Clifton prefers to describe this narrative as a "fabrication."

His own version of this standard narrative is given in his chapter entitled "The Indian Story: A Cultural Fiction" (1990:29-47), and is worth quoting at length in order to get the flavour of it:

> In the beginning, North America was motherland for between ten and thirty million truly humane beings. This dense population was organized into over two hundred separate, sovereign nations existing continuously–according to the unquestionable authority of their own traditional histories–from time immemorial. Each such sovereignty had its own government and exclusive national territory. Although none of these indigenous nations understood or recognized the propriety of owning, buying, or selling land, they did claim and exercise the rightful privilege of occupying parts of it and using its fruits. This right, as hosts, they freely shared with their neighbors and visiting strangers, whom they treated generously as guests...Each with its own language and special customs marking their unique identities, these nations lived in peace and harmony

> with one another. Underneath these minor cultural differences, none-the-
> less, lay vitalizing commonalities, the heart and soul of the Indian. Each
> nation, for instance, defined its territory as a Holy Land, and all together
> they worshipfully personified their habitat as Mother Earth, existing in
> harmony with all her creatures. This biological or environmentalist ethic
> pervaded every aspect of the life of the Indian, for whom all things, all
> thought, all behavior, and all happenings were pervasively sacred.
> Animistic, purely spiritualistic, uncontaminated, these archaic nations
> existed in free-floating, ahistorical time, their beliefs and ways
> irreversible, insoluble, and–as others have but recently come to
> appreciate–ineradicable (1990:32-33).

This gives the general flavour of the narrative which Clifton describes, which he claims is "in whole or substantial part fictitious" (1990:41). Clearly, some aspects of the narrative Clifton describes *are* fictions. The claims that all Native Americans lived in harmony without wars between them, that they existed as they are presently constituted since time immemorial, and that their traditions are static, timeless and unchanging are all easily falsified by the historical, ethnographic and archaeological records. Other aspects of the narrative were not true of *all* Native American societies, such as the proposal that they were all egalitarian (as shall be discussed below).

Just as clear, however, is the fact that many aspects of the narrative are *not* fictions. Much of it is historical fact. Other images seem to be an attempt to describe the common themes which best characterize Native American philosophies and lifestyles–especially as they are distinct from those which are characteristic of the dominant Western culture.

The present section shall discuss two of these images or philosophical themes in more detail. Both images are highlighted by Clifton, who describes them as examples of "The Good Things the Indian Gave the Whiteman" (1990:23), presumably in order to suggest that these images are just as false as the utopian claim that there were no wars among the various First Nations. The first is the idea of the "Noble Savage," or of using the Indian as a model for "an egalitarian ethic," which has a long history in Western political philosophy. The second, the Indian as a model for "ecological sainthood"–while also being connected to the image of the Noble Savage–has gained political importance in more recent times, especially with the rise of the ecological movement in the past few decades.

Besides discussing the question of whether each of these images provides a more or less accurate picture of Aboriginal philosophies and lifestyles, the present section shall also discuss the political role which each image has played, and continues to play, in the Western world. For as shall be argued below, each theme provides an example of ways in which Native American philosophies appear to have entered into, and changed (or begun to change) the direction of Western patterns of thought.

Yet while both points of view are clearly out of step with the dominant patterns of practice and belief of Western culture prior to the colonial encounter (and, arguably, still are), both are quite commonly found in accounts of Native American views from the earliest recorded accounts up until the present day. The influences are also largely the result of the *example* which their ways of life provided. This constrasts markedly with attempts to *coerce* others into accepting their views through force of arms or forced assimilation, which has been the common Western pattern of influence on Native cultures. The contrast between these two patterns of influence also reemphasizes the difference between the *egalitarian* and *hierarchical* patterns of social organization most typical of each type of culture.

Each theme also leads directly into a discussion of the social and ecological patterns of relationship enacted by the two cultural types, and to the central theme of this section–an abductive comparison of these patterns to the relational symbolisms discussed above. In doing so, it shall continue to turn primarily to Native American literature for support in its attempts to interpret Native American philosophy, and to trace the history of this philosophy as far back as the available literature can take us. In other words, the discussion shall continue to refer primarily to Native American understandings of themselves when describing their philosophies. This is, after all, a courtesy which we routinely apply to our academic colleagues when tracing the history of ideas.

2. THE TWO FREEDOMS

> I said to General Howard...We are all sprung from a woman, although we
> are unlike in many things. We cannot be made over again. You are as you
> were made, and as you were made you can remain. We are just as we were
> made by the Great Spirit, and you can not change us; then why should
> children of one mother and one father quarrel–why should one try to cheat
> the other? I do not believe that the Great Spirit Chief gave one kind of
> men the right to tell another kind of men what they must do.
> General Howard replied: You deny my authority, do you? (Chief Joseph, as
> cited in: Vanderwerth 1971:268).

> Before the pale faces came among us, we enjoyed the happiness of unbound-
> ed freedom, and were acquainted with neither riches, wants nor oppression.
> How is it now? Wants and oppression are our lot; for are we not controlled in
> everything, and dare we move without asking, by your leave? Are we not
> being stripped day by day of the little that remains of our ancient liberty?
> (Tecumseh, as cited in: Vanderwerth 1971:63-64)[48].

THIS CHAPTER shall deal with two related questions. The first is
the whether all Native American cultures were uniformly
egalitarian. Despite the sentiments expressed by Tecumseh, above,
we shall see that the answer to this question is not an unequivocal
"yes." The second question is the role which Western images of
Native Americans as models for more egalitarian politics has
played, nonetheless, in Western political philosophy. For despite
the superficial or distorted views of Aboriginal cultures held by
some Western thinkers, this image played an important role
historically, and continues to in the present day.

Concerning the question of whether Native American cultures
were egalitarian, the answer is a simple, "Yes and no." This is
because there is abundant ethnohistorical, ethnographic and archaeological evidence to suggest that many Native American polities
were actually ranked societies, or chiefdoms, which exhibited clear
differences in status. Donald (1990), in his essay in Clifton's

---

[48] Tecumseh was speaking in 1811 to a gathering of Choctaws and Chickasaws,
attempting to convince them to join his war against the United States.

collection, reviews the evidence and suggests that there are two cultural areas north of Meso-America which were distinctly *not* egalitarian.

The first examples are the cultures of the Southeastern United States, east of the Mississippi River (1990:150-53). In this area there appears to be clear evidence, both ethnohistorical and archaeological, for the existence of complex, ranked chiefdoms. As he suggests, "we have considerable archaeological evidence from Southeastern sites to suggest that at various times and places some of the region's prehistoric communities contained socially stratified populations" (1990:151). Two of the classic examples of this type of society are provided by Cahokia and Moundville (Price and Feinman 1993:236-67). The clearest evidence of ranking is found in differences in the richness of grave goods found in ancient burials, which were common to both of the examples mentioned. A common assumption among archaeologists, as Donald suggests, is that where such differences in richness of burials crosscut both gender and age differences, then there is strong evidence for the inheritance of status or rank within families; that such status is ascribed to particular lineages, rather than being achieved through one's individual efforts as in egalitarian societies.

Historically, Donald also points out two cultures which appear to have been ranked into at least two classes. The first is the Calusa of southern Florida, and the second is the Natchez, who were destroyed by the French around 1730 (1990:152).

The other culture area which has exhibited ranking historically, are the cultures of the Northwest Coast. As Donald suggests, "in terms of inequality, all followed a basic pattern" (1990:153). In general, these societies were characterized by three ranks, or strata: "Titleholders" or nobles, commoners and slaves. Not only were these cultures internally ranked between nobles, who inherited their positions, and commoners, but also between the members of the local group and slaves taken in raids and wars with other nearby groups. These slaves were slaves in the truest sense: they were considered to be the property of particular nobles, they were required to work for them, and they could be killed at the whim of their owners.

Thus, there appears to be abundant evidence to support the fact that *all* Native American societies were not egalitarian. Yet as Donald himself points out, the aim of his essay "is not to show that there were no genuinely egalitarian societies...in Native North America. Surely there were" (1990:149). This is not what one might assume from reading Clifton's introduction to the collection in which his essay appears. For part of the narrative which Clifton proposed, and which he considers to be "in whole or substantial part fictitious," reads as follows:

> Freely given cooperation was the norm in all things political and economic, made possible because everyone owned all necessary means of production...the evils of political, economic, social, or gender inequality were unknown. Political hierarchy was incomprehensible... (1990:33).

This is clearly a description of classic egalitarian societies, yet as shall be illustrated below, this image is not uniformly false for all Native American cultures. To my mind, Murray Bookchin has best encompassed the differences between egalitarianism—as understood by many Native American thinkers—and the "equality" which hierarchically organized Western democracies practice. His contrast is between "the inequality of equals" and "the equality of unequals," or between *formal* equality and *substantive* equality, respectively (1991:140-66).

The inequality of equals describes the pattern of the Western world. Here everyone is *considered* to be "equal before the law," despite the fact that they are grossly *unequal* in wealth, talents and social power. "Equality," in this sense, is primarily focused upon the *relationship between individuals and the state*, and the ideal is that all citizens be treated equally by the state, in spite of their in-dividual differences. For example, a rich businessman and a beggar in the street may both receive the same fine for the same crime, despite the fact that this sum may represent pocket change to the one, and more money than they have seen for several months to the other. And a stint in jail awaits the beggar who cannot pay. Thus, they are considered to be equals even though they are not, and in a society in which some are born to riches, and others to poverty, there is even talk of "equal opportunity for all."

Following social archetypes, the pattern here is hierarchical, and people are free to *compete* with one another for *unequal* positions of wealth and privilege within this social hierarchy. The fact that people are *born into* unequal positions within this hierarchy is most often conveniently ignored. The pattern which has resulted is an increasingly centralized pattern of economic control, which is little different from that of the Middle Ages in some respects. For if the common (and convenient) separation of political from economic forms of rule is dissolved, it becomes quite evident that the majority of the richest members of Western culture have inherited their positions of wealth and influence in precisely the same manner in which Medieval monarchs did.

The equality of unequals, on the other hand, describes the pattern which was typical of many First Nations, and of other egalitarian peoples throughout the world. Here the inequalities in talents and abilities which people naturally have are accepted as given, yet every effort is made to bring about an *equal distribution of property*, and of the essential supports of life. This type of egalitarian philosophy is closely tied to a pattern of generalized reciprocity within the group as well. As Standing Bear pointed out in 1933, for example:

> The central aim of the Lakota code was to bring ease and comfort in equal measure to all. There were no weak and no strong individuals from the standpoint of possessing human rights. It was every person's duty to see that the right of every other person to eat and be clothed was respected and there was no more question about it than there was about the free and ungoverned use of sunshine, pure air, and the rains with which they bathed their bodies. There were no groups of strength allied against groups of those weak in power (1978:123-24).

In other words, there were no *classes*, and the Lakota, at least, seem to have recognized an "irreducible minimum" to which all were equally entitled, which is a common sentiment throughout the Native literature[49]. Those who had the ability to acquire more goods through hunting or gathering or horticulture or crafts were expected to help those who were less fortunate. Indeed, it was often considered to be "a great honour to be asked by those in want to share your food, clothing, horses, or any comfort of life" (Standing Bear 1978:28). Or as Eastman puts it, "[t]he true Indian sets no price upon either his property or his labour. His generosity is limited only by his strength and ability" (1980:103).

So Marx's famous dictum, "from each according to ability, to each according to need" does describe this reciprocal pattern of relations rather well, at least among one's own people[50]. As has often been said, the chief of a people often appeared to be the poorest of the group, since maintaining the respect of the people required that they live up to these ideals, and lead by the example which they provided. Here people are considered to be *unequal* in

---

[49] For a discussion of the prevalence of the irreducible minimum concept among tribal peoples more generally see Bodley (2001:194-7).

[50] Needless to say, relations between groups, such as trading, did not always follow such a reciprocal pattern.

ability, yet to enjoy an *equal right to a decent life*, and every effort is made to bring about a more equitable distribution of the necessary supports for life. As Harold Cardinal, Cree once encapsulated the difference between these two views of equality–the formal and the substantive:

> Theoretically, of course, all Canadians have equal status, even Indians. Yet any fool can read government figures showing that one in five Canadians is poor enough to be in a state of dependency (1969:133).

So where Western democracy seems to represent the freedom of individuals to compete for *unequal* positions of wealth and influence, egalitarianism in Native American philosophies seems to represent the freedom to *cooperate* in order to secure a good life–not only for oneself, but for the group as a whole. Once again, egalitarianism is closely related to the pattern of reciprocity which many Aboriginal groups practiced (including even ranked societies, though in a form through which the central redistributor reaffirmed his hereditary status through his generosity). For what one gives to another today, when one has a surplus, will be returned at another time when one is in need, just as in the contemporary saying: "What goes around, comes around." This also frees one from worrying about starving to death when there is more than enough to go around, for as Standing Bear observes, "hunger is a hard thing to bear, but not so hard when all are sharing the same want in the same degree; but it is doubly hard to bear when all about is plenty which the hungry dare not touch" (1978:163). Or as Sitting Bull once expressed this same difference in outlook, "The white man knows how to make everything, but he does not know how to distribute it" (as cited in Brown 1970:427).

Where in the one view the economic and political leaders gain respect due to their positions of power and influence within the social hierarchy, in the other one gains positions of influence due to respect for one's example, knowledge, skill and generosity. For in the political organization of many First Nations:

> Men in council were there because of merit. A man might be poor in goods, own few horses, and live in a small tipi, but he would sit with the council. Riches bought no man power and though he might have many horses he could not buy a seat with the wise ones (Standing Bear 1978:25).

So while the latter point of view recognizes that some people are better than others, both morally and in the practice of various skills, *this* sense of superiority does *not* imply that they are justified in arbitrarily ruling over their inferiors. Nor are the interests of the superior maximized at the expense of their fellows. As Standing Bear elaborates upon the views of his Lakota people:

> the council made no laws that were enforceable upon individuals. Were it decided to move camp, the decision was compulsory upon no one...But it must not be thought that there was no distinction among Lakotas as individuals. There were no social strata so definite that some were unattainable by reason of class or birth, but there were individuals and groups who were recognized by virtue of superior intelligence and capacity. As among every people, there were those who were better able to get along in the world...So there were the high–those who were honoured and distinguished for superior achievement. There were others who, though not excelling, were good members of society. There were still others who, through no fault of their own, found themselves a little weaker, less capacitated, than the majority members of society. These were not the old and the children, but adults, able-bodied, but unable to achieve. This being the case, nevertheless, weaklings were never the objects of pity, charity, or contempt...They were never allowed to want, nor to grow sick and die from neglect. But in spite of the careful equality of treatment, these three distinctions were kept in mind. The Lakotas of high class, if such they may be called, were proud, to be sure, but it was this pride that would not permit them to overlook want and suffering in their kind. The people of this class did not believe in waiting to be called upon to give aid, but took pride in seeing that they were never asked (1978:129-30).

What Standing Bear's passage suggests is that chiefs did not so much *rule* the people through coercion, but rather *lead* the people by persuasion and example. Indeed, leadership roles were often transitory and task-oriented in egalitarian societies, or of short duration and tied to particular tasks. Thus, when a war party returned home, or a hunt or ritual was successfully completed, the "authority" of those who lead the particular activity came to an end, until next they initiated such an action, or were called upon by others to do so. This implies that to raise a war party, for example, chiefs or leading warriors had to convince others that the actions were necessary and important, and as Wooden Leg suggests, "Ordinarily the advice of the chiefs was heeded. But the obedience was a voluntary one. In battle, the chiefs had no authority to issue commands that must be obeyed" (1931:120).

While differences in wealth and ability, or between superior and inferior persons were recognized, and the superior were held up as models for the emulation of their fellows, the assumption that their superiority justified their right to rule arbitrarily seems to have been conspicuously absent. Indeed, the influence of the leaders of many Native American groups seems not to have been based upon command and obedience so much as upon simple respect for their knowledge, their skills and their persons. Thus, their influence arose more spontaneously from the general will of the people, and was not imposed from a position of assumed superiority. This seems to be the key difference between egalitarian and hierarchical patterns of social and political organization–as symbolized by the circle and the pyramid in the previous section.

As my reference to Bookchin has already obliquely illustrated, Native American societies have often served as important models or examples in Western political philosophy. Indeed, it appears likely that the egalitarian polities which existed historically in North America served as important models for the advent of democracy, and of relatively more egalitarian views, in the political philosophy and rhetoric of the Western world as well (if not so fully in its practice). As Jack Weatherford proposes:

> Egalitarian democracy and liberty as we know them today owe little to Europe. They are not Greco-Roman derivatives somehow revived by the French in the eighteenth century. They entered modern western thought as American Indian notions translated into European language and culture (1988:128).

This brings us to the second question with which I began this section, for Weatherford suggests that it was only with the advent of the colonial encounter that these relatively egalitarian ideas began to enter into European thought. True, the *word* "democracy" was derived from ancient Greek, but democracy as we understand it today was anything but a Greek idea. Neither in the writings of Plato nor Aristotle, nor in the actual practice of what the Greeks called "democracy" did the ideals of freedom and equality come to the fore. Quite consistent with their opinions, only the adult male citizens of Athens could vote. Their wives, their slaves, and resident aliens (many of whom were born in the city) were denied the franchise. According to some estimates only about 15% of the population of Athens had the right to vote (Plato 1974).

The concept of a "freeman" seems also to have had none of the more egalitarian implications we might like to give it today in the ancient Greek literature. It seems to have meant something along the lines of "not a slave," or "able to rule himself;" that is, over slaves, wife and children if not the state. Aristotle considered that one could be a freeman even where there was only a single ruler, and thus saw no contradiction between freedom and monarchy. Indeed, for Aristotle there was no essential connection between freedom and democracy at all. Being a freeman simply implied that one was a *person* and not *property*, and must be treated as such. As he put it himself, "a slave is a living tool and a tool is a lifeless slave," and "[t]he use made of slaves...departs but little from that made of other animals" (1986:518 and 552). Presumably the slaves were simply more easily trained than other domestic animals, and at a greater variety of tasks, but this makes them more valuable only in an instrumental sense, and not in a moral one[51]. As Jean-Jacques Rousseau was later to characterize this argument:

> Just as the shepherd is superior in kind to his sheep, so, too, the shepherds of men, or, in other words, their rulers, are superior in kind to their peoples. This, according to Philo, was the argument advanced by Caligula, the Emperor, who drew from the analogy the perfectly true conclusion that either Kings are Gods or their subjects brute beasts (Barker 1960:171).

This is the reason why the distinction between a freeman and a slave as different *in kind* was so important to the arguments of Aristotle. For as he stated, "the rule over freemen differs from that over slaves no less than men who are free by nature differ from men who are slaves by nature" (1989:595). This results in two types of rule appropriate to each; "political rule" and "despotic rule."

The rule of master over slave is considered equivalent to despotic rule, and is in the interest *of the ruler only*. The rule over freemen, or political rule, should be *in their own interest*, or what he describes as the "common interest"–despite the fact that the interests of slaves, women and resident aliens are excluded from consideration. Yet either the rule of the "majority" in a democracy, the rule of the "few" in an aristocracy, or the rule of a single man in a kingdom could be termed "political rule." The ruler simply had to rule *in the interests of freemen* (and since he considered the interests of freemen to be different in kind from those of slaves, slavery was not inconsistent with his position). Yet while Aristotle recognized that monarchy was the most risky form of political rule, since its

---

[51] As Maybury-Lewis points out, Aristotle's doctrine of "natural slavery" was often used as a justification for the subjugation of non-Western peoples in the early colonial age as well (2002:11).

deviant form of despotic rule, or "tyranny," was the worst of the despotic forms, his ideal type of government remained the rule by a single enlightened ruler.

Similarly, when Plato (1974) set out to construct an ideal society in *The Republic*, he constructed an unmistakably class based society. Many of the ideas which might be considered egalitarian today–such as equality for women and the elimination of extremes of wealth and property–were meant to apply only to the ruling class, or the Guardians. The ideal was also the rule by a single Philo-sopher King–chosen from among this Guardian class–while the common people, and the slaves, were to carry on as usual. So it seems that equality also had different connotations than it is often given today among the ancient Greeks, for it was only a restricted sphere which was considered equal. As Aristotle related, "it is thought by some men that what is just is the equal; and it is the equal, but for persons who are equal and not for everybody" (1986:576). Thus hierarchy remains the prevalent ideal, and it seems that slaves, as well as women and foreigners, did not fit into the category of "persons" at all.

That being the case, one is forced to look elsewhere for the true origins of democratic ideals, for little changed throughout the Middle Ages. This is why Weatherford suggests that it was only with the advent of the colonial encounter that egalitarian ideals began to enter into European thought. Not surprisingly, he traces the origins of these ideals to European contacts with the very "uncivilized" types of peoples whom Aristotle would have described as "incapable of rule," in this case to Native Americans.

As Weatherford points out, Sir Thomas More's *Utopia* was not completed until 1516, almost two decades after Columbus' voyage to America, and the discovery of peoples who apparently lived without government in the European sense (1988:122). Letters and accounts by explorers such as Amerigo Vespucci were also much discussed among the educated Europeans of the time, and More's vision–of a utopia without money, where all persons were equal–was both highly original in the European tradition, as well as similar to these accounts in the above respects.

All early accounts of Native American politics were not uni-formly positive. Nor did they all emphasize the egalitarian or "utopian" nature of indigenous societies as a model for Western society. Neither did they all turn to indigenous philosophies and lifestyles in order to critique the shortcomings of their own society's political organization. Over a century after More's work appeared, for example, Thomas Hobbes published his classic in political philosophy, *Leviathan*, which argued in support of the English monarchy of the time.

Suddenly–and not coincidentally, soon after the discovery of egalitarian cultures by European explorers and traders–it became necessary to explain why humanity had submitted to the authority which had previously been taken for granted in European culture, and the myth of the "social contract" was born. Hobbes argued that people entered into the social contract, and created a common ruler over all, because life in the "state of nature" which preceded "civilized" society was a state of perpetual warfare; a war of each against all in which "the life of man [was] solitary, poore, nasty, brutish, and short" (1985:186). What often goes unmentioned is that the state of chaos he described was based upon his conceptions of the life of Native Americans, however misguided they may have been. As Hobbes pointed out himself:

> It may peradventure be thought, there was never such a time, nor condition of warre as this; and I believe it was never generally so, over all the world: but there are many places, where they live so now. For the savage people

in many places of *America,* except the government of small Families, the concord whereof dependeth upon natural lust, have no government at all; and live at this day in that brutish manner, as I said before (1985:187).

Where according to Aristotle the uncivilized barbarians were not properly human, for Hobbes an egalitarian way of life was not considered to be a society. Instead, it was equated with the state of chaos (nature) which he described. Consequently, "society" was also unapologetically equated with centralized government, and with the institutions of social control *within* society, or with the hierarchical organization of state level societies. Just as with current political rhetoric, anything else was dismissed as "anarchy" or "chaos." In this way any form of organization other than the hierarchical may be dismissed out of hand[52].

The writings of John Locke, appearing in 1690, modified Hobbes' position slightly. Instead of dismissing aboriginal life as a chaotic state of war between "everyman and everyman," Locke proposes that; "The natural liberty of man is to be free from any superior power on earth, and not to be under the will or legislative authority of men, but to have only the law of Nature for his rule" (1960:15). Though Locke argues that "in the beginning all the world was America," he still attempts to explain, and thereby to justify, the origins of civil governments, which are again equated with "society." Once again, it is also the insecurity of one's person and, significantly, one's *property* in the state of nature which he sees as driving people into a social contract, and making them give up their natural freedom for the lesser freedom of life under a civil government.

Locke attempts to justify hierarchy by pointing out that one's

---

[52] In fact, anarchism simply means lack of hierarchical rule, rather than lack of order, and of all Western political philosophies it is the theory which is closest to what the majority of First Nations actually practiced.

excess possessions would have no protection in a state of nature, since to build up large amounts of private property a government is required, and "government has no other end but the pre-servation of property" (1960:55). Its origins, according to Locke, lie in the advent of individual greed for more possessions than one requires for subsistence, and a desire to keep them for oneself even if one has no immediate need for them.

Locke himself appears to assume that this was always what humans desired—that the desire to accumulate and maximize personal wealth is equivalent to human nature. He fails to explain, for example, how the desire to accumulate large amounts of private property managed to arise from the pattern of reciprocity and redistribution of possessions which was actually typical of most of the Aboriginal peoples whom he would classify as living in his "state of nature." Yet as Eastman related in 1911, "It was our belief that the love of possessions is a weakness to be overcome" (1980:99). Or as George described the ways of his people, in a com-ment echoing many others found throughout Native American literature:

> My culture did not price the hoarding of private possessions, in fact, to hoard was a shameful thing to do among my people. The Indian looked on all things in nature as belonging to him and he expected to share them with others and to take only what he needed (1974:40-1)[53].

So it seems that Locke's assumptions concerning human nature are far from being universal, and stand in need of revision. Yet without these assumptions, like Hobbes before him, Locke can offer no reason to explain why people would voluntarily enter into a "social contract" in the first place. In this case the more likely explanation for the advent of hierarchical organization, or Locke's "civil society," is that the vast majority of people did *not* enter into it

---

[53] A similar sentiment is expressed by Standing Bear (1978:168) .

voluntarily. Rather, as was the case in the colonial age, it is more likely that it was *imposed* upon them by force of arms.

Yet a later entry into the social contract debate, who is even more concerned with liberty and the equality of all persons, is also more sympathetic with what he construes to be Native American views–this being Jean-Jacques Rousseau. For while Rousseau, not unlike Hobbes, still held to the untenable hypothesis that humans had originally been solitary beings, coming together only to reproduce, he felt that this state had disappeared in very ancient times, and had preceded the state of nature in which Native Americans and other Aboriginal peoples were living in his day.

Native ways of life also became the ideal for Rousseau, from which hierarchical society was an unhappy decline. As evidence he cited the fact that many "savage" peoples of his day were still reluctant to give up their natural freedom for "civilized" life, even after centuries of attempts by missionaries, traders and many others to convince them of the benefits of civilization (1984:168-70)[54].

There were also many accounts of Native American views circulating in Europe at the time. Rousseau himself was almost certainly aware of the writings of Baron de Lahontan, in whose works is found the following quotation attributed to Adario or Kondiaronk, a Huron chief with whom he spoke circa 1700:

> In earnest, my dear Brother, Im sorry for thee from the bottom of my soul. Take my advice and turn HURON; for I see plainly a vast difference between thy condition and mine. I am Master of my Condition and mine. I am Master of my Body, I have the absolute disposal of my self, I do what I please, I am the first and last of my Nation, I fear no Man, and I depend only on the Great Spirit. Whereas, thy Body, as well as thy Soul, are doomed to dependence upon thy great Captain, thy Vice-Roy disposes of thee, thou hast not the liberty of doing what thou hast a mind to...and

---

[54] Bodley (1999) makes a similar observation concerning tribal peoples in the Twentieth century, which shall be discussed in more detail in Section IV.

though dependest upon an infinity of Persons whose Places have raised them above thee. Is it true or not (as cited in McLuhan 1971:50)[55]?

As this succession of social contract thinkers illustrates—all of whom were attempting to explain why the Western world no longer lived as Native Americans were observed to live—the more concerned they become with liberty and equality, the more sympathetically they treat Native ideas, until they come to be known as the "noble savage." And as Weatherford relates:

> The concept of the 'noble savage' derived largely from writings about the American Indians, and even though the picture grew romanticized and distorted, the writers were only romanticizing and distorting something that really did exist. The Indians did live in a fairly democratic condition, they were egalitarian, and they did live in greater harmony with nature (1988:129).

True, the precise wording of Baron de Lahontan's translation, and others of the period, may have been tailored somewhat to suit the political philosophies of various European scholars, but the ideas expressed were firmly rooted in fact. Sioui, a Huron scholar himself, provides a detailed analysis of the dialogue between Lahontan and Adario entitled "Lahontan: Discoverer of Americity" (1992:61-81). Yet while he admits that the dialogue is contrived, rather than an actual quotation, he argues for the essential accuracy of Lahontan's account of Wendat philosophy.

There is also no denying that Lahontan, and others of the time, looked to Native American political organization and ideals as a model for their own philosophies—just as Rousseau did—however accurately or inaccurately they may have rendered them in their

[55] While Donald proposes that Adario was "probably fictional," and that the dialogue with Adario was contrived (1990:146), other historians do not agree. McLuhan, from whom the quotation is cited, suggests that Adario "played a prominent part in Frontenac's War" (1971:50). Similarly, the Wendat or Huron scholar Sioui affirms the actual existence of Adario, stating that "For the Wendat, Adario simultaneously assumed the functions of chief of 'war'...and council chief" (1992:66).

writings. Moreover, the popularity of the idea of the noble savage and the politics which went along with it just prior to the French Revolution is surely no coincidence. For it is the writings of Rousseau himself which are generally considered to be one of the key intellectual influences which fueled the political discontent of the French, and led to the overthrow of the French royalty.

In America, a similar role was played by the writings of Thomas Paine, whose pamphlet *Common Sense* is considered largely responsible for turning the American revolution into a war for freedom and independence from Britain, rather than for re-cognition as citizens of its empire. As Weatherford points out, both Paine and Benjamin Franklin were also well acquainted with the political organization of such nations as the Iroquois. Franklin, for example, was present when the following words were spoken by Canasatego, an Onondaga *sachem*. Speaking at a treaty conference in Pennsylvania in 1744, where Franklin was recording the con-versations, it was Canasatego who originally suggested that the English colonists form a union similar to that of the Iroquois Confederacy:

> We heartily recommend Union and good agreement between you, our [English] brethren...Our wise forefathers established union and amity between the Five Nations; this has made us formidable; this has given us great weight and authority with our neighboring nations. We are a powerful Confederacy; and, by your observing the same methods our wise forefathers have taken, you will acquire fresh strength and power (1992:116).

Franklin not only suggested that this course be taken, but that the newly forming United States should adopt many other features directly from the organization of the Iroquois League[56]. So as the council of the Six Nations Iroquois Confederacy itself claims:

---

[56] For more information on the ideas which the United States actually adopted from the Iroquois and other Native Americans see Weatherford (1988:117-31).

European people left our council fires and journeyed forth into the world to spread principles of justice and democracy which they learned from us and which have had profound effects upon the evolution of the Modern World (Knudtson and Suzuki 1992:192)[57].

If any doubt remains that egalitarian ideals are a common theme in early Native oratory and writings, I hope that the following quotation will serve to dispel them. Published as "An Indian's Views of Indian Affairs," it originally appeared in the *North American Review* in its April 1879 issue. Though this is long after democracy of a type was established in the West, one must remember that the person speaking could not read or write English. Thus he had no knowledge of Western political philosophy beyond what he had seen in practice, or heard about from those he knew. It is instructive that he does not see democracy as he understands it in those practices, and is worth quoting at length. In the words of Chief Joseph of the Nez Perce First Nation, who was speaking in Washington, D. C . on January 14, 1879[58], shortly after being defeated in war, having his lands taken from himself and his people by force of arms, and being confined to a reservation:

If the white man wants to live in peace with the Indians he can live in peace. There need be no trouble. Treat all men alike. Give them all the same law. Give them all an even chance to live and grow. All men were made by the Great Spirit Chief. They are all brothers. The earth is the mother of all people, and all people should have equal rights upon it. You might as well expect the rivers to run backward as that any man who was born a free man should be contented when penned up and denied liberty to go

---

[57] The quote is taken from a declaration by the Iroquois passed on April 17, 1979, and later published in Akwesasne Notes in its Spring 1979 issue. As Knudtson and Suzuki point out, "The council has requested that their declaration be given the widest circulation possible" (1992:218).

[58] Interestingly, Sam Gill, whose arguments shall be discussed below, suggests that this speech was recorded in Oklahoma, where Joseph was confined, in order "to record his recollections and views" (1987:51-2). This despite the fact that Joseph himself notes in the speech in question that he was delivering it in Washington, as when he states that "At last I was granted permission to come to Washington...I am glad we came" (as cited in Vanderwerth 1971:281).

where he pleases...I have asked some of the great white chiefs where they get the authority to say to the Indian that he shall stay in one place, while he sees white men going where they please. They cannot tell me. I only ask the government to be treated as all other men are treated...We only ask an even chance to live as other men live. We ask to be recognized as men. We ask that the same law shall work alike on all men. If the Indian breaks the law, punish him by the law. If the white man breaks the law, punish him also. Let me be a free man—free to travel, free to stop, free to work, free to choose my own teachers, free to follow the religion of my fathers, free to think and talk and act for myself—and I will obey every law or submit to the penalty. Whenever the white man treats an Indian as they treat each other, then we will have no more wars (Vanderwerth 1971:282-83)[59].

Since Chief Joseph's eloquent plea for freedom and equality was addressed to an audience which included many of the congressmen, cabinet members, and diplomats of a government which claimed these ideals for its own, it is evident that the terms were given a different meaning by the two. This difference reflects the contrast between egalitarian and hierarchical political systems. To Chief Joseph, the government of the United States appeared to be assuming a type of authority over others which belonged, properly, only to the Great Spirit. As he stated at another time, "Say to us if you can say it, that you were sent by the Creative Power to talk to us. Perhaps you think the Creator sent you here to dispose of us as you see fit. If I thought you were sent by the Creator I might be induced to think you had a right to dispose of me" (as cited in Brown 1970:316).

But then, as was pointed out above, throughout the Middle

---

[59] In his narrative, Joseph comments that he came to Washington [D C] with his friend Yellow Bull "and our interpreter" (Vanderwerth 1971:281). Though he does not give the name of this person, apparently it was someone whom he trusted to interpret his words truthfully and well. Though excerpts from the speech are widely available, the complete speech is worth reading. Against the backdrop of his account of the circumstances which lead up to the war on his people, and of the war itself, his words are among the most eloquent pleas for freedom and equality which I have ever had the pleasure to read. The entire speech is reproduced in Vanderwerth (1971:260-84).

Ages and earlier, hierarchy was often justified by a priesthood which claimed to be God's representatives on Earth, who were thus–literally–playing God. God Himself was also conceived of after the model of an Earthly King. The position of which he accuses them, therefore, was not inconsistent with their heritage.

It seems evident, however, that the first steps towards eroding this hierarchical heritage in the West–with the advent of a form of centralized, representative (rather than participatory) democracy, and egalitarian ideals and rhetoric (if not practices)–were taken as a result of Western attempts to assimilate the fact that egalitarianism itself, particularly as practiced and advocated by various Native American peoples, *was an actual possibility*. Both the image and the reality of egalitarianism as a central theme in Native American philosophies and polities, therefore, have long been central to the political understandings of the West.

The principle difference between these two understandings of freedom and equality appears to be that the Western world sees no contradiction between its own centralized, hierarchical pattern of organization and the egalitarian principles it so recently adopted. As with Hobbes and Locke, egalitarian patterns of organization such as those of various First Nations are often not even considered to be a society, let alone a polity. If one looks carefully at the manner in which the political options are represented by contemporary political rhetoric, political scientists and the news media, one sees that little has changed. For the usual symbolism is that of left, right and center, with the left being more concerned with the rights of the working class, and the right being more concerned with the interests of the capitalist class. One should note, however, that the only kind of society which can be encompassed by this classificatory system is a hierarchically organized one.

Egalitarian societies, such as many First Nations historically were, cannot even be fit on the map, and are therefore not an option.

From a cross-cultural perspective, however, this system is revealed as totally inadequate for dealing with the various forms of political organization which people have, in fact, practiced. All of the political forms which can be fit onto the map share a common pattern of hierarchical relationships. In this respect, at least, they are more similar than different. Yet their opposite–egality–which becomes quite evident in light of cross-cultural comparisons, is excluded from consideration here.

Thus, behind the initial appearance of a symbolism of balance lies the usual pattern of opposition of the dominant Western relational symbolism. Once again we return to a symbolism which separates and opposes dualisms, considering one extreme to be the ideal, and denigrating or dismissing the other. In this case the dismissal has been so effective in practice that many contemporary Westerners do not fully comprehend that people have ever lived in any other manner than the hierarchical ideal which government, the media, and much of academia promote. Even those who are aware of itgenerally dismiss the possibility of an egalitarian pattern of social relations as "utopian," despite the fact that people have, in fact, successfully lived according to more egalitarian patterns of life throughout the majority of human history[60].

These two ways of understanding the relationship between opposites are also evident in the practices of the two democracies. The egalitarian democracy of many First Nations was most typically a small scale, participatory democracy. Where decisions affecting the entire group were thought to be necessary, they would be made in open counsel, so that opinions could be openly expressed and

---

[60] Including contemporary Western subcultures such as the Quakers, the Amish or the Hutterites.

discussed face-to-face. Native decision making was also largely based upon *consensus*, where a general agreement was reached after lengthy discussion.

As Basil Johnston describes the deliberations of his Anishinabe people, "There was no debate. Instead the speakers sought illumination through mutual inquiry" (1982:171). The style was not confrontational. Consequently, presenting one's views in a way which everyone could agree with was important for a leader. This required a balanced understanding of the views of the entire group, and a balancing of individual opinions with the needs of the group as a whole. And where consensus could *not* be reached an opinion of part of the group, even if that of the majority, could not be imposed upon the rest of the group.

In a representative democracy, on the other hand, it is the majority which rules, so that the will of the majority is maximized, and is often imposed upon an unwilling minority. Yet one must keep in mind that the rule of the majority can be understood in two senses. The first is the majority of the voting public. Usually it is only once every several years that they are allowed to participate in the functioning of government by electing a new batch of officials to "represent" their views. Even then the public does not vote directly upon policy, but only for the people who shall decide it. It is largely the established political parties, and not the general public, which select the candidates among whom they may choose, and the issues to be debated as well. The second tier in this hierarchy is a majority of the elected representatives, and once again the views of the minority may be largely ignored. This is a pattern which lends itself to competitiveness and factionalism.

Without coercive institutions with which to enforce the will of the majority upon an unwilling minority, this was not usually an option in egalitarian politics, where a more cooperative pattern prevailed. In any case, the minority could simply leave if they disagreed with the majority's decision[61]. Whether the will of a majority of elected officials is truly representative of the will of a majority of the people is also open for debate case by case. They make up only a fraction of one percent of the population, and as always in a hierarchical pattern of organization a minority, in fact, decides for the majority.

Rousseau, for one, felt that such representation of the will of the people was a chimera, and criticized the emerging representative democracy in the England of his day with words which he would likely have applied to the practices of contemporary Western democracies if he were alive today:

> Sovereignty...consists essentially of the general will, and will cannot be represented...The Deputies of the People are not, nor can they be, its representatives. They can only be its Commissioners. They can make no definitive decisions. Laws which the People have not ratified in their own person are null and void. That is to say, they are not laws at all. The English people think that they are free, but in this belief they are profoundly wrong. They are free only when they are electing members of Parliament. Once the election has been completed, they revert to a condition of slavery: they are nothing. Making such use of it in the few short moments of their freedom, they deserve to lose it (Barker 1960:260).

The dominant Western political philosophy has also had a tendency to separate and oppose society and the individual. This is true, for example, of the opposition between liberal individualism and communism, which was so typical of Cold War rhetoric. For the two were thought to be mutually exclusive, with the former being most concerned with the individual, and the latter most concerned with the society. And though both patterns have been dominant in different parts of the West historically, they shared a

---

[61] This process is formally referred to as group fissioning in the anthropological literature.

basic relational symbolism. For either option excluded the other as inferior and extolled their chosen side of the duality as superior, and both share a hierarchical pattern of relations.

It is surely true that Native Americans tended to place more emphasis upon the group, and less upon the individual, than is typical of atomistic liberal philosophies. As with the social contract thinkers described above, it was the individual, in the West, which was considered to be basic. The advent of society required an explaination.

As Lame Deer contrasted Native views with the extreme individualism of the West, "We aren't divided up into neat little families—Pa, Ma, kids, and to hell with everybody else. The whole damn tribe was one big family; that's our kind of reality" (1972:34). Or as Okute of the Teton Sioux once stated, again with reference to natural archetypes (and the importance of the natural diversity emphasized by Charles Darwin):

> Animals and plants are taught by *Wakan Tanka* what they are to do. *Wakan Tanka* teaches the birds to make nests, yet the nests of all birds are not alike. *Wakan Tanka* gives them merely the outline. Some make better nests than others. In the same way some animals are satisfied with very rough dwellings, while others make attractive places to live. Some animals also take better care of their young than others...All birds, even those of the same species, are not alike, and it is the same with animals and with human beings. The reason *Wakan Tanka* does not make two birds, or animals, or human beings exactly alike is because each was placed here by *Wakan Tanka* to be an independent individuality and to rely on itself...From my boyhood I have observed leaves, trees, and grass, and I have never found two alike. They may have a general likeness, but on examination I have found that they differ slightly. Plants are of different families...It is the same with animals...It is the same with human beings; there is some place which is best adapted to each (McLuhan 1971:18-19)[62].

In Native American patterns of thought, then, the individual

---

[62] Most of this passage is also cited by Deloria (1994:88-89), in support of his own discussion contrasting Native American religions with Christian conceptions. It was originally published in Densmore (1918). Okute, or Shooter, was an elder at the time, and was speaking in 1911.

and the larger group are both realities, with neither emphasized at the other's expense. Rather, one would expect that the balancing of these opposites, or the establishment of an harmonious pattern of relations between them, would be the ideal. Only according to a logic which separates and opposes dualities are the two seen as necessarily opposed to one another, with a gain for society being a necessary loss for individuals, and vice versa.

Once again, this way of thinking is partly the result of equating society with the institutions of social control *within* society, and of following social archetypes and the hierarchical logic which they suggest. For indeed, in *this* sense of society the two *are* often opposed, since the laws passed by such hierarchical social institutions often *do* infringe upon the freedom of the individuals which comprise the *actual* society. Indeed, as Chief Tony Mercredi encapsulated this difference in political philosophy, echoing my discussion of relational symbolism above:

> Envision, if you will, a circle. The Creator occupies the centre of the circle and society...revolves around the Creator. This system is not based on hierarchy. Rather, it is based on harmony. Harmony between the elements, between and within ourselves and within our relationship with the Creator. In this circle there are only equals. Now, envision a triangle. This triangle represents the fundamental elements of the Euro-Canadian society. Authority emanates from the top and filters down to the bottom. Those at the bottom are accountable to those at the top, that is control. Control in this society is not self-imposed, but rather exercised by those at the top upon those beneath them. In this system the place of the First Nations peoples is at the bottom. This is alien to the fundamental elements of our society, where we are accountable only to the Creator, our own consciences and to the maintenance of harmony (RCAP 1997)[63].

---

[63] Quoted from Vol.2, Part IV: "Lands and Resources," Chapter 3.1 "Lessons from the Hearings." Mercredi was then Chief of the Athabasca Chipewyan First Nation, and was speaking to the RCAP at Fort Chipewyan, Alberta, on June 18, 1992.

3. ON MOTHER EARTH AND THE CONSERVATION ETHIC

we will not be denied the privilege without molestation of visiting at any time the tombs of our ancestors, friends and children. Every part of this soil is sacred in the estimation of my people. Every hillside, every valley, every plain and grove, has been hallowed by some sad or happy event in days long vanished. Even the rocks, which seem to be dumb and dead as they swelter in the sun along the silent shore, thrill with memories of stirring events connected with the lives of my people, and the very dust upon which you now stand responds more lovingly to their footsteps than to yours, because it is rich with the blood of our ancestors and our bare feet are conscious of the sympathetic touch (Chief Seattle, as cited in: Vander-werth 1971:121)[64].

I think this is one feature that most Indians have in common. They have a deep attachment for the land (Bennie Bearskin, as cited in: Nabokov 1991:349).

THE PRESENT SECTION shall deal with the idea that Native Americans provide a model for "ecological sainthood," which Clifton claims is "in whole or substantial part fictitious." While the popular image of Native American peoples as "original conservationists" has become quite common, especially as the settler culture begins to struggle with ecological issues, it is not without its detractors.

Sam Gill (1987, 1990), for example, has argued that the concept of Mother Earth among Native Americans was actually borrowed from the Western world, and has arisen only within about the last 150 years. As he states it, "[t]he story of Mother Earth begins almost concurrently with the story told by Smohalla in 1885" (1990:131). Smohalla belonged to a band of Nez Perce living in

---

[64] Seattle's speech has generated some controversy, due to the fact that a fictional version of the speech attributed to him, which was actually a movie script, was widely quoted by the ecology movement for a time. The entire history of the episode, as well as all three different versions of the speech, is available in Gifford and Cook (1992). The above quote is from the original translation, recorded in 1854 by Dr. Henry Smith, and first published in the Seattle Morning Star, October 29, 1887.

eastern Washington state, and was the leader of a revitalization movement known as the "dreamers." This movement encouraged its followers to reject the ways of the colonizing culture, and was especially opposed to the practice of agriculture. As Smohalla put it in the often cited quotation:

> My young men shall never work. Men who work cannot dream, and wisdom comes in dreams...You ask me to plow the ground? Shall I take a knife and tear my mothers bosom? Then when I die she will not take me to her bosom to rest. You ask me to dig for stone! Shall I dig under her skin for her bones? Then when I die I cannot enter her body to be born again. You ask me to cut grass and make hay and sell it, and be rich like white me, but how dare I cut off my mother's hair? (as cited in Gill 1990:131).

Gill then claims that all other references to "Mother Earth" in the Native literature of this period can be linked to a knowledge of Smohalla and his movement. He also claims that an earlier reference attributed to Tecumseh, the Shawnee leader, is entirely fictitious (1987:26).

Methodologically, Gill is a particularist, who emphasizes the diversity characterizing Native American religions over their commonalities. He prefers to see and to highlight differences rather than similarities. As he states, "I find it hard to think of religion and culture in ways that are not rather firmly grounded in the particular" (1987:157). Or again, "I have found no basis for conceiving of the religious traditions of the hundreds of tribes native to North America for a single, homogeneous tradition" (1987:3). And thus Gill criticizes a long lineage of scholars not only because they cite so few examples in support of their arguments for the "existence" of Mother Earth, but also, implicitly, because they are generalists. That is, they attempt to understand the common themes of Native philosophies or religions more generally (as do the Pan-Indian movement and the present work).

Given the link between the image of Mother Earth and a

conservation ethic in contemporary Aboriginal oratory and literature, one might be tempted to assume that Gill has demonstrated
that a conservation ethic was also adopted from the West. This is
not the case. In fact, what Gill was looking for in his inquiry was a
goddess figure, specifically named "Mother Earth," and rituals
surrounding her, which he was unable to find. The existence of the
conservation ethic which is often expressed *through the use of this
metaphor* was not even considered, though Gill did admit that "The
earth is spoken of as a mother in metaphorical terms" (1987:41).
Thus, whether Gill's arguments concerning the goddess Mother
Earth are valid or not, they do not prove that an ethic of respect for
the land did not exist long prior to Smohalla's statement. That
Tecumseh himself expressed such views is illustrated by the
following reference to the manner in which European settlers
treated the land, and its political implications are clear:

> Where today are the Pequod? Where are the Narragansetts, the Mo-
> hawks, the Pocanokets, and many other once powerful tribes of our people?
> They have vanished before the avarice and oppression of the white man,
> as snow before a summer sun...Look abroad over their once beautiful country,
> and what see you? Naught but the ravages of the pale face destroyers meet
> our eyes. So it will be with you Chocktaws and Chickasaws! \ Soon your
> mighty forest trees, under the shade of whose wide spreading branches you
> have played in infancy, sported in boyhood, and now rest your wearied
> limbs after the fatigue of the chase, will be cut down to fence in the land
> which the white intruders dare to call their own. Soon their broad roads
> will pass over the graves of your fathers, and the place of their rest will be
> blotted out forever (Vanderwerth 1971:63).

In this speech, given in 1811, Tecumseh clearly emphasizes to
his listeners not only the way in which they are likely to be treated
in future by the settler population, but also the destruction of their
lands and forests. He also characterizes settler populations as
"destroyers" not only for their reputation as destroyers of Indians,
but *of their lands* as well. This speech also emphasizes one important

reason why many Native Americans seem to have considered the lands to be sacred–because they contained the bones, blood or graves of their ancestors. This is a common expression of respect for the land, which is found in several early Native speeches, some of which predate Smohalla's speech by several decades[65].

The aboriginality of a "conservation ethic," or an ethic of respect for the land, particularly among the Cree of the Hudson's Bay drainage area, has also been called into question by Brightman (1993). The Cree provide a particularly useful case study due to the fact that there is not full consensus among anthropologists who have studied them concerning the question of whether the Cree traditionally had a conservation ethic or ideology–and especially concerning its origins–even though all admit that respect for the land and animals is a common sentiment in recent times.

Studies of Cree resource management in the present century uniformly reveal a variety of conservation practices in their patterns of hunting, fishing and trapping, as well as the existence of a conservation ideology and rituals surrounding it (Berkes 1998, 1999; Berkes et al 1994, 1995; Brightman 1993, Feit 1973; Speck 1977; Tanner 1979). This ideology of respect for the animals hunted was first documented by Speck, based upon his studies among the Montagnais-Naskapi, who live in the boreal forests of Quebec. Speck characterized this attitude as "a feeling that they owe a debt to the animal world for its sacrifice of life on their behalf" (1977:74). He also claimed that there had been little change in these attitudes since they were first documented by French priests in the seventeenth century (1977:73). Such attitudes have also been documented throughout the Eastern James Bay Cree (Preston 2002),

---

[65] See, for example, the speeches made by Black Hawk of the Sauk in 1832 (Vanderwerth 1971:86-87), and Chief Seattle of the Suquamish and Duwamish tribes of the Washington area in 1854 (Vanderwerth 1971:121).

including the Waswanipi (Feit 1973), Mistassini (Tanner 1979), and Chisasibi Cree (Berkes 1998, 1999; Niezen 1998) in more recent studies, all of whom live in the eastern boreal forest region which Speck discussed.

Both Martin (1978) and Brightman (1993), however, have used an ethnohistorical approach to question whether such an ethic existed during the early fur trade in different ways. Martin, for example, argued that epidemics of European diseases were blamed upon animals by various Algonquian peoples, and that this lead to a "war against the animals" which they carried out through the fur trade. The central argument of his thesis–that boreal forest Algonquians such as the Cree associated disease sanctions with animals–however, has largely been discredited (Krech 1981).

Brightman, on the other hand, argues that the Cree of the eighteenth and nineteenth centuries in the area southwest of Hudson's Bay had neither an ideology, nor a practice of conservation. Instead, he suggests that they learned their conservation practices from the Hudson's Bay Company (HBC) in historical times, in response to an episode of over-hunting. There are two main lines of evidence which Brightman presents in order to argue these points, both of which are closely related.

The first argument suggests that an *intensification* of Cree hunting and trapping in the areas southwest of Hudson's Bay in the late eighteenth and early nineteenth centuries caused widespread depletions of both large game, as well as animals trapped so that their furs could be traded, such as beaver. As Brightman states, "the expansion of inland trading posts prompted intensified rates of predation by Crees which led to long-term decline in beaver, moose, and caribou populations" (1993:292). While he presents good evidence for widespread game depletions after about 1810 (1993:266-68), which have also been documented by others (Ray 1974), his central thesis that *intensified* hunting on the part of the Cree was the *primary cause* of these depletions is questionable for several reasons.

To begin with, Brightman presents no positive evidence from Hudson's Bay Company records that an intensification of hunting and trapping actually took place. In other words, he does not provide positive evidence showing that the company's annual take in furs actually increased during this period. Rather, he *assumes* that there was such an intensification in order to explain the game depletions themselves, and that this intensified hunting was their primary cause. In doing so, he neglects to consider other possible causes in any detail. As he states, the boreal forest ecosystem is characterized by low species diversity and by "extreme fluctuations in animal populations" (1993:245), and these fluctuations are caused "by many factors unrelated to human predation" (1993:286). Yet while he admits that intensified hunting may have interacted with other factors, such as disease (1993:272), he does not explore these possibilities in any detail.

Arthur Ray, to whom Brightman refers concerning the latter point, *does* explore these possibilities. Ray provides two quotations from the early 1800s which describe an epidemic which destroyed beaver populations in many areas of western Canada, particularly those which "lived in the ponds and stagnant water" rather than in areas of running water, which were less affected (1974:119). Ray also points out that periodic droughts occurred during this period, and falling water levels would have impacted muskrat and beaver populations. Further, the drought led to forest fires, which were "common" in this period as well. Such fires would not only have affected the populations of species such as moose and other big game–both by directly killing them and by destroying their browse–but were said to have "killed large numbers of beaver" (1974:120) as well.

Ray also proposes that the big game shortages were not as extensive as Brightman claims, suggesting that "[g]ame conditions in the lands adjacent to Saskatchewan were apparently much more favourable" (1974:122), particularly for the big game species. Thus, while he admits that hunting and trapping played a role in these depletions–as is intuitively obvious–Ray also emphasizes the importance of other "cyclic variations" and "natural calamities" in the depletion of the species upon which the Cree relied for the bulk of their subsistence (1974:121). Further, Ray does not suggest that trapping was the *primary* cause of the depletions, as Brightman does, nor does he argue for an *intensification* of these activities.

A further ecological factor which Ray highlights–in the area of human demography–may explain why. This is that two major epidemics occurred in the area which Brightman discusses in the decades prior to the major game depletions. The first was an epidemic of smallpox in 1780, which sources at the time claimed had killed from "one-half to three-fifths" of the Natives affected (1974:105-6). This was followed by a dual epidemic of measles and whooping cough, which swept throughout the Western Interior in 1819 (1974:106-7), resulting in population losses of 40% or "even greater" (1974:108). Thus, even though Ray documents populations rebounding fairly quickly for some groups in the area, a *decrease* of this magnitude in the number of available hunters and trappers during the period leading up to the game depletions is unlikely to have lead to an *intensification* of trapping[66].

---

[66] The depopulation does, however, appear to have preceeded, and perhaps even prompted, the expansion of HBC posts inland, in an attempt to maintain their quotas by involving more distant bands in the fur trade.

Finally, the intensification hypothesis is questionable given the information on Cree attitudes which Brightman himself provides, but does not choose to emphasize. This is that Cree hunting and trapping was based upon a principle of "least effort" (1993:247), or upon "a disposition toward satisfying finite needs with little labour" (1993:261). Like many other foraging peoples, the Cree seem to have recognized a principle of *sufficiency* (Sahlins 1972). One illustration of this is the fact that "Crees *reduced* the number of furs they brought to the posts when offered better prices since it then took fewer furs to obtain the goods they required" (Bright-man 1993:249-50, emphasis added). Brightman also provides evidence that the traders themselves often recorded their continual exhortations to the Indians "not to be Lassy," due to the fact that the supply of furs provided by Native peoples was not keeping up with European demand, and suggests that such statements provide "little evidence for unremitting Indian industry" (1993:251).

This demonstrates not only that Cree foraging strategies were guided by a very different ethic from that of the capitalist West–which did not seek to maximize returns–but that their desire for European technology and trade goods was limited (1993:251), and did not lead to an intensification of hunting and trapping activity in and of itself. Rather, when they had enough furs to trade for what they needed for household self-sufficiency, they simply stopped trapping and hunted big game for meat. This would have "rested" the beaver and other fur bearers while focusing upon moose, caribou and other big game species.

Consequently, Brightman is forced to provide a different rationale for the game depletions, and for the intensification of trapping and hunting which he claims. His answer is that they arose from indigenous conceptions of human-animal relationships.

More specifically, Brightman argues that the Cree practiced "in-discriminate killing" (1993:297) of any animals they encountered, whether they needed the meat or not, because they conceived of game populations as "infinitely renewable" (1993:288). He then cites a variety of early sources which claim that the Cree routinely killed many more animals than they could consume, and made only selective use of the carcasses. A source from 1733 from the Hudson's Bay post describes Cree beliefs and practices as follows:

> they have a maxim very prejudicial to this country which is that the more beasts they kill, the more they increase; and in consequence of this they destroy great numbers for the sake of the tongues, leaving the carcasses to rot (Brightman 1993:287).

Brightman claims that this doctrine was a result of Cree conceptions of animal "bosses" or "owners" which controlled the abundance and movements of animals, and directed them to be taken by the hunters (1993:288). Consequently, "if a hunter is 'offered' much by these spirits, it is obligatory to take much. Indiscriminate killing...discharges the obligation to receive" (1993:290). Brightman argues that the Cree "'respected' animals in religious terms while unknowingly decimating them...the dominant ideology of human-animal relationships not only allowed indiscriminate hunting but enjoined it" (1993:290).

This line of evidence is also debatable, however, for two reasons. The first is the nature of the evidence for this belief. It is supplied exclusively by Hudson's Bay Company employees, who did not travel with the Indians when they hunted, but only observed their practices when they were near the post. Thus, their descriptions of Cree beliefs could easily have misrepresented what the Cree they spoke with were actually attempting to say when describing their hunting practices, since most of their knowledge was based upon what they heard, rather than upon what they observed. Brightman also admits that there was "enthusiastic plagiarism" among the early chroniclers of the Hudson's Bay Company, and thus that "these accounts need not all be taken as independent attestations" (1993:288).

Indeed, many of these accounts were also descriptions of the so-called "Home Guard" Cree, with whose hunting practices the Hudson's Bay Company employees were most directly familiar. The reason was that the "home guard" remained near the post throughout the year, rather than hunting elsewhere. Besides sporadic employment, an important part of the reason for this was that many wished to remain close to the post in order to trade meat and furs for alcohol. Sometimes they would even kill animals at times when the post had no need of them in an attempt to procure it. Perhaps this particular group was more concerned with assuring a continuous supply of alcohol, rather than with assuring a continuous supply of animals, and as Brightman admits, "it would be possible to interpret them as a breakdown in aboriginal patterns" (1993:255-6).

Clearly it would, yet this is something which Brightman fails to do, which raises a second reason that this line of argument is questionable. For when Brightman discusses contemporary Cree conservation practices, he is careful to emphasize two things. First, that there is a great *diversity* in the practices of individual hunters and trappers, both in their actual *practices* and in their *explanations* for those practices (1993:317). Second, he explains that the concern for conservation *fluctuates*, depending upon the abundance of particular species, and documents how it has done so in the present century (1993:314-15). When populations of beaver are high, little conservation is practiced, while when populations are low, various measures are taken to conserve their numbers–especially more selective killing.

Yet Brightman does not consider that both of these circumstances may also have been true historically. He notes both that there are scattered references to "beaver management" (1993:258) and early evidence of "selective trapping" (1993:290). Yet the former evidence is considered exceptional, while he describes the latter as "a subdominant and superficially Western doctrine" (1993:290). Instead of emphasizing the possible diversity which characterized Cree beliefs and practices in the past, Brightman chooses to emphasize the universality of those beliefs and practices which are consistent with his hypothesis of intensified hunting, while de-emphasizing the fact that "some groups may have conserved animals...while others did not" (1993:259).

Neither does he consider that in the past, as in the present, practices may have responded to changes in animal populations, with conservation being practiced more diligently in times of scarcity–such as after the game depletions discussed around 1810–and less diligently in times of abundance. Instead, Brightman claims that if such conservation practices "were widespread, either pre-historically or in the fur trade or both, the beaver depletions strongly suggest that they were subsequently abandoned during the late eighteenth and early nineteenth century" (1993:259). Thus, his selection of evidence seems tailored to confirm the hypothesis discussed above, that game depletions were primarily the result of over-hunting. Yet his hypothesis is also questionable for a variety of reasons, and he provides little direct evidence to substantiate it.

While all agree that hunting strategies have changed and evolved in response to the fur trade, concerning the aboriginality of a conservation ethic among the Cree there are two possibilities, depending upon how one interprets the evidence. On the one hand, such conservation practices may have a recent origin, being either independently invented, or learned from Hudson's Bay Company employees, as Brightman claims. After all, Brightman does provide clear evidence that many *Hudson's Bay Company employees* conceived of Cree hunting as based upon the idea that animals were "infinitely renewable resources whose numbers could neither be reduced by over killing nor managed by selective hunting" (1993:280). He also provides evidence that the Hudson's Bay Company attempted to introduce various measures intended to conserve beaver populations after the collapse.

In this case, both the practice and the ideology of conservation are a recent innovation, and it would appear that Cree notions of "respect" for animals have evolved in response to a major game depletion and/or in their interaction with their European inter-locutors. This view assumes that originally, the Cree world view did not associate hunting with depletion. Because boreal forest fauna populations fluctuate widely on their own, and the Cree were highly mobile—having not yet evolved their hunting territory system—they may have lacked the necessary feedbacks to inform them of the effects of their hunting and trapping activities (Berkes 1998:120). From this perspective it is only when they receive obvious feedbacks from their own over-hunting that a conserv-ation ideology evolves. In other words, "Now that hunters know that it is possible to deplete game animals by over-hunting and that wastage does matter, their value systems change accordingly" (Berkes 1988:123). This view assumes that with pre-contact techno-logy the Cree had no knowledge of the possibility of over-hunting, or that Cree patterns of exploitation and movement left them in a position in which they were *unaware* that it was a possibility.

The other possibility is that the historical sources Brightman cites, and Brightman himself, have misinterpreted Cree ideology, and that a conservation ethic is in fact aboriginal. This is consistent with Cree oral traditions and anthropological observations since early in the twentieth century (Speck 1977). For while there is clear historical evidence that *some* Crees were over-hunting, it is an assumption to claim that *all* were, and that this ideology did not exist. After all, cases of over-hunting have been documented in the past century even for people who *do* have a conservation ethic.

The caribou in the Chisasibi area were over-hunted early in the twentieth century, for example, and again when they initially returned to the area some 70 years later. Significantly, however, it was the conservation ethic of the elders, conveyed through stories of the earlier episode of over-hunting, which allowed the practice of conservation to be restored when the caribou returned (Berkes 1999:95-109). This illustrates the truism that cultural ideologies and cultural practices are not always consistent. For as Berkes suggests, "[a]nyone who has worked with hunting peoples knows that rules of ethics are sometimes suspended. But one can say that about any culture...there is always a gap between the ideal practice and the actual" (1999:95).

The above episode also illustrates the process of cultural learning through which this ethic has expressed itself historically. And this type of diversity, and the vacillation between conservation and non-conservation in times of scarcity and abundance, were very likely characteristic of the Cree of the eighteenth and nineteenth centuries as well, though we lack first hand accounts of their views from that time to confirm or deny this. Perhaps this conservation ethic has been learned–and relearned–for a much longer period of time, and repeatedly transmitted through the traditional ecological knowledge of the more respected and experienced hunters and trappers to their cohorts in times of resource crisis.

In either case, there is no debate concerning the existence of a Cree conservation ideology, and Cree conservation practices from the twentieth century onwards. Further, Morantz' (1986) review of HBC records concerning the James Bay area from the eighteenth century onwards–a region which Brightman did not study–reveals that conservation was not mentioned until the early nineteenth century, "and then is discussed as though such practices were well entrenched." As he concludes, "[t]his suggests that conservation measures had already been observed for some time" (1986:86). It also suggests either that Brightman's thesis is incorrect, or again highlights the internal diversity characterizing Cree hunting and trapping strategies in different areas.

Not unlike Tanner's (1986:32) assertion concerning the origins of family hunting territories, it might be more useful to set aside the question of whether a conservation ethic is *aboriginal* in favour of the question of whether it is *Algonquian*. In other words, like the family territory system, it is not so much a question of whether a conservation ethic existed prior to contact, but whether it evolved *internally*, either in response to the fur trade, or from tendencies which already existed within pre-contact Cree culture.

The real question is whether a conservation ethic was of indigenous origin, or whether it was learned from contact with, the settler culture. Given the fact that Western culture has only begun to grapple with issues of conservation and sustainability in the last few decades, while the earliest Western expression of an explicitly ecological ideology of respect for the land can be traced to Aldo Leopold's (1949) writings, the latter option appears highly unlikely. If anything, the pattern of influence has been in precisely the opposite direction.

4. MASTERS AND CONQUERORS OF NATURE

> At the bottom of everything, I...continue to believe, is a religious view of
> the world that seeks to locate our species within the fabric of life. As long
> as Indians exist there will be conflict between the tribes and any group that
> carelessly despoils the land and the life it supports (Deloria 1994:1).

THIS BRINGS ME to the second significant influence which Native
American philosophies have had, or could have, upon Western
philosophy. This is also one of the central themes of the present
work: the origins of ecological ideals. This theme is also closely
related to the discussion of political philosophy in Section II:2, for
the domination of other peoples or other persons is logically
connected with the domination of nature by humanity in Western
philosophy. As in the theory, so also in the practice. For it was not
only First Nations which had to be conquered and subdued, but
nature itself. Both the "wild man" and the "wilderness" stood in the
way of civilization's advance.

In Native philosophies, on the other hand, egalitarian respect
for other persons tends to be connected with a more respectful
attitude towards nature. As Young Chief of the Cayuse expressed
such a sentiment in 1885, "The Great Spirit, in placing men on the
earth, desired for them to take good care of the ground and to do
each other no harm" (as cited in McLuhan 1971:8).

Just as the relational symbolism of the ideological sphere is
reflective of the patterns of relationship in the social sphere in each
view, so also these social patterns of relationship are carried into
the ecological sphere by which they are subsumed (whether or not
they are consistent with the patterns of relationship of the ecological sphere itself). As an overview reveals, the three form a
more-or-less self-consistent whole in each of the broad cultural
types which are being considered here–those of the Western and
Native American nations. As in the ideological and social spheres,

their respective views of the land, and their ways of relating to it, also differ in several fundamental ways. One of the classic illustrations of this remains the different attitudes towards wildlife, and especially the buffalo, which are reflected in the way in which each of the two treated them. As Neihardt interpreted Black Elk's sentiments concerning the differences between the two:

> The *Wasichus* did not kill them to eat; they killed them for the metal that makes them crazy, and they took only the hides to sell. Sometimes they did not even take the hides, only the tongues...You can see that the men who did this were crazy. Sometimes they did not even take the tongues; they just killed and killed because they liked to do that. When we hunted bison, we took only what we needed (Black Elk 1988:213)[67].

These sentiments echo those of Satanta, a Chief of the Kiowas, who once compared the behavior of the whites to that of ignorant children, who seemed unaware of the most basic rules of survival: "Has the white man become a child that he should recklessly kill and not eat? When the red men slay game, they do so that they may live and not starve" (as cited in Brown 1970:241). Bear Tooth of the Crows was equally shocked, suggesting that it was not only the buffalo which were the victims of this irresponsible attitude: "Fathers, your young men have devastated my country and killed my animals, the elk, the deer, the antelope, my buffalo. They do not kill them to eat them; they leave them to rot where they fall" (as cited in Brown 1970:144). Aleck Paul of the Chippewa noted the same fundamental difference between the attitudes of Natives and whites, and like Black Elk, traced their lack of respect for the value of animals to the one thing which they seemed to value above all

---

[67] A pertinent section of the original interview notes reads: "The four-leggeds and the wings of the air and the mother earth were supposed to be relative-like and all three of us lived together on mother earth—we all had teamwork at that time. Because of living together like relatives, we were doing just fine. We roamed the wild countries and in them there was plenty and we were never in want. Of course at that time we did not know what money was" (DeMallie 1985:288).

else. As he stated, "[w]hen the white people came they commenced killing all the game. They left nothing on purpose to breed and keep up the supply, because the white man don't care about the animals. They are after the money" (as cited in Nabokov 1991:86). Thus, as Cajete concludes:

> Eighteenth century Native people must have been shocked by Western hunters' appalling slaughter and maiming of Plains buffalo, not only because they depended so heavily upon the great beasts, but also because they had never experienced an entire culture with no respect for the equality and spirit of animals (2000:175).

Despite their lack of knowledge of Western philosophical literature, their reading of it from an observation of the broader practices of the settler culture was again particularly insightful. For in the writings of John Locke is found a consideration of the very differences in value which each attaches to the land, and that which lives on it, to which they allude. Locke also draws a distinction between the monetary value of things, and "the intrinsic value of things, which depends only on their usefulness to the life of man" (1960:23). Locke felt that in the "state of nature" the animals taken in the hunt were valuable due to the fact that by eating them, the hunter and his people might continue to live. In "civil society" they were valuable only due to the profit which could be received by exploiting them. Locke proposed a "labour theory of value" to explain the difference between the two, in which "'tis labour indeed which puts the difference of value on everything," while "nature and the earth furnished only the almost worthless material as in themselves."

Thus, nature is considered to be worthless unless it is made use of by humanity for some economic purpose, and has little if any monetary value until it is exploited by humanity. Or as Standing Bear suggests, "To the Lakota the magnificent forests and splendid herds were incomparable in value. To the white man everything was valueless except the gold in the hills" (1978:44).

The natural intrinsic value of nature–as a support for life–is separated and opposed to the social value which is attached to nature by money–its value as a commodity on the market. The latter type of value arises from the labour which is mixed with nature as it is transformed into a marketable commodity, which can only be produced if it is appropriated from its natural state. Needless to say, it is the latter type of value which is considered to be superior, and which must be maximized, while the intrinsic value of nature is dismissed, and replaced by the value of money. Once again, it is also the "desire of having more than man needed" which is seen as having "altered the intrinsic value of things." So from the desire to maximize one's personal possessions, the denigration of the intrinsic value of nature follows.

In support of his labour theory of value, and the superiority of the monetary value which flows from that labour, Locke proposes that "[t]here cannot be a clearer demonstration of anything than several nations of the Americans are of this." For while Native lands were just as fruitful as those of England, because they did not "improve" it through their labour, "if all the profit an Indian received from it were to be valued and sold here it would not be worth one thousandth as much." Thus Locke argued that:

> land that is left wholly to nature, that hath no improvement of pasturage, tillage, or planting, is called, as indeed it is, *waste*; and we shall find that the benefit of it amount to little more than nothing (1960:25-26. emphasis added).

The benefit which he described as "little more than nothing" included the lives of the Indians, since they depended upon that which the land naturally offered up for their subsistence–that is, upon its intrinsic value as a support of life–even without the benefit of "improvement." Evidently the Indians themselves were also considered to be next to worthless, since they produced little if anything of economic value, which Locke considered to be the superior standard of valuation. So as Eastman once observed:

> The Native American has been generally despised by his white conquerors for his poverty and simplicity. They forget, perhaps, that his religion forbade the accumulation of wealth and the enjoyment of luxury. To him, as to other single-minded men in every age and race...the love of possessions has appeared a snare, and the burdens of a complex society a source of needless peril and temptation...It was not, then, wholly from ignorance or improvidence that he failed to establish permanent towns and to develop a material civilization (1980:9-10).

Thus, the fact that First Nations did not develop a technological civilization after the Western model was not due to ignorance, but due also to the fact that their ethics and ideologies did not tend to value such a development. So while Locke denigrated the intrinsic value of things, in identifying this form of valuation with Native American philosophies he did provide a fairly accurate account of their views. For they do seem to have valued nature, and life, for their own sake, and not merely as a means of producing a profit. As Lame Deer relates, "To us life, all life, is sacred, since All of nature is in me, and a bit of myself is in all of nature" (1972:111 and 126). Yet unlike Locke's account, in Native philosophies the intrinsic value of nature seems to have depended upon a thing's usefulness to life in general, or to life itself, rather than narrowly "to the life of man." The term "life" also encompassed everything, including that which the Western world would consider to be inanimate. Unlike the human-centered definition of the West, theirs is a far more inclusive understanding of the idea of intrinsic value. As Standing Bear once put it:

> We are of the soil and the soil is of us. We loved the birds and beasts that grew with us on this soil. They drank the same water as we did and

breathed the same air. We are all one in nature. Believing so, there was in our hearts a great peace and a welling kindness for all living, growing things (1978:45).

Due to the egalitarian and participatory views expressed in Native American philosophies, humanity is considered to be a part of the Earth, intimately connected to all other aspects of nature. The differences between humans, animals, plants, and soil are not considered to imply superiority and inferiority. Nor do they consider themselves justified in exploiting nature due to its presumed inferiority. They did not kill animals and plants because they were inferior and existed merely for the use of humanity, but because—as is natural—they *must* if they wished to continue to live. For life and death are inseparable, since every time that we eat another organism inevitably dies so that our own life may be sustained.

They did not separate and oppose humanity and nature, nor consider themselves to be inherently superior to nature. Rather, since they considered themselves to be a part of nature, they seem to have felt that any value which they attached to their own lives also had to be extended to encompass the natural conditions which made those lives possible—their "selves" were not *separate* from nature. And if the people wished to survive, a balance had to be struck between the needs of the people, and the needs of the game animals and plants which they required for that survival. Once again they follow natural archetypes, and a natural model of that which is valuable, which results in a view in which nature itself is afforded ethical regard.

Cajete describes this philosophy as "natural democracy" (2000:52). Rather than seeing it as an extension of the *social* ideals of egalitarianism and reciprocity to encompass nature, he also proposes the opposite, that "nature was the primary model for Native community" (2000:104). This theme seems to be one of the defining features of a Native American point of view, which is essential to an adequate understanding of their world view and ethos. This is the sentiment of respect for the Great Spirit through respect for the world it continually creates through its immanent presence in all things–in all people, and in all of nature. Since this sacred power moves in and through all, animating and giving life to all of nature, turning to nature as a model after which to patterns one's own styles of life and thought is to model them after the divine. All of nature must be respected if the Great Spirit itself is to be respected. Or as it was expressed in an anonymously authored pamphlet I received from a Native American friend, this type of "Nature Mysticism" is "aware of the presence of the Great Spirit through-out the entire cosmic order." As a consequence:

> From the very first, the Native Indian has centered his life in the Natural World. He is deeply invested in the earth–committed to it in his consciousness and in his instinct. Only in reference to the earth can he persist in his true identity. This is why the Native Indian conceives of himself in terms of the land. In his view the earth is sacred. It is a living thing in which living entities have origin and destiny–the Native Indian is bound to the earth in his spirit. By means of his involvement in the Natural World does the Indian ensure his own well-being[68].

Thus life itself is the basic value, a value which is considered to be intrinsic to all living things. The Earth itself is also considered to be a living entity, which provides the source from which all other beings and all other values flow. For Locke, however, as for the cultures which he represents, "the earth and all inferior creatures" exist merely to be used. Humanity is considered to be separate from and superior to nature–and because of this superiority, nature exists only to be exploited by humanity. Indeed, it is the

---

[68] My copy of "Our Way" was provided to my by Neil Hall, a Native drummer and singer, and then a resident of Winnipeg, Manitoba, when I met with him to discuss the meaning of contemporary powwow gatherings in November 1992.

very criterion by which a thing's "usefulness" is to be judged which has been reinterpreted by Locke, and by his "civil society." Its value is to be judged according to its usefulness in producing monetary values, or profit, rather than in perpetuating life. The evaluation of a thing's usefulness is arrived at in abstraction from its usefulness to the perpetuation of life, and from nature, which has been reduced to nothing more than a collection of "natural resources." These are valuable only contingent upon their being used in the production of marketable commodities, and thus in the production of monetary values.

It follows from Locke's view that because Native Americans do not know how to "properly" exploit their land, in order to maximize the amount of monetary value which can be derived from it, it is considered to be "waste." And a better justification for stealing their lands would be difficult to find, for such waste was surely near criminal to the emerging capitalist mentality of Europe at the time when Locke was writing. The similarity of his reasoning to the doctrine of Manifest Destiny should also be apparent. As a defense of this doctrine penned in 1870 reads:

> The rich and beautiful valleys of Wyoming are destined for the occupancy and sustenance of the Anglo-Saxon race. The wealth that for untold ages has lain hidden beneath the snow-capped summits of our mountains has been placed there by Providence to reward the brave spirits whose lot it is to compose the advance-guard of civilization. The Indians must stand aside or be overwhelmed by the ever advancing and ever increasing tide of emigration. The destiny of the aboriginies is written in characters not to be mistaken. The same inscrutable Arbiter that decreed the downfall of Rome has pronounced the doom of extinction upon the red men of America (Brown 1970:189)[69].

As with Locke's philosophy, because First Nations did not "properly" exploit their lands by maximizing the monetary value of

---

[69] Significantly, these words were written by a group of white frontiersmen who wished to open the area to their mining expedition. Originally published in the Cheyenne Daily Leader, March 3, 1870.

its natural resources, they must stand aside so that those who knew "better" than they could put it to "better" use. As Locke phrased it, their land was considered to be "more than the people who dwell on it, do, or can make use of." And the standard of value by which he arrives at this position is based entirely upon social archetypes. For as Locke himself points out, the value attached to gold or diamonds is purely conventional, while they have no intrinsic value whatever.

It is only by means of maximizing economic value that civilization may "advance" in such a view. Even today "progress" and economic expansion remain closely linked as concepts. And it was in the name of progress that the buffalo were exterminated. As General Sheridan remarked at the time; "Let them kill, skin, and sell until the buffalo is exterminated, as it is the only way to bring lasting peace and allow civilization to advance" (as cited in Brown 1970:265). The buffalo stood in the way of those who wished to ranch and farm the plains, and had to be exterminated in order to maximize the economic gain which could be realized from these activities. Their extermination would also eliminate the chief source of food and provision of the plains tribes, and thus the war on the buffalo lent itself to the subduing of both nature and the First Nations of the plains as well.

As Chief Joseph once observed, "We were contented to let things remain as the Great Spirit Chief made them. They were not; and would change the rivers and the mountains if they did not suit them" (as cited in Brown 1970:316). From a Native American perspective, humans were considered to be *a part* of the land upon which they lived, and were meant to adapt themselves to the ways of nature, or to fit themselves intelligently into the natural patterns of relationship which were observable there.

For the Natives themselves, this identification with the land was seen as a positive and harmonious thing. Only in the eyes of the Western world was transformed into a negative. Though the West also identified Native peoples with nature, as they did themselves, the result was that the negative evaluation associated with all things natural came to be applied to the Natives as well. Indeed, they often received little better treatment than the buffalo themselves. As Standing Bear recalls the time in which the buffalo were slaughtered, which he witnessed in his youth:

> The Indians were never such wasteful wanton killers of this noble game animal...I saw the bodies of hundreds of dead buffalo lying about, just wasting, and the odor was terrible...we saw bale after bale of buffalo skins, all packed, ready for market. These people were taking away the source of clothing and lodges that had been provided for us by our Creator, and they were letting our food lie on the plains to rot. They were to receive money for all this, while the Indians were to receive only abuse...These people cared nothing for us, and it meant nothing to them to take our lives, even through starvation and cold. This was the beginning of our hatred for the white people (1975:67-8).

So the domination of nature through the killing of the buffalo was inextricably linked to the domination of the Indian peoples themselves. Indeed, in Canada the plains tribes were forced into treaties without a shot being fired, through the simple expedient of destroying the buffalo, and with it their supply of food and provision. As Joseph Dion relates, after a smallpox epidemic in 1871-1872, and a general famine for several years due to the extermination of the buffalo, his Plains Cree people were eager for relief at any price. As he states:

> Times were then very hard; the buffalo were almost extinct and the Crees were trying desperately to uphold their former pride and dignity, but nevertheless often felt the pangs of hunger...The people cared little and knew nothing of the enormity of the deal and what they were relinquishing in exchange" (1979:76).

In other words, it is very easy to convince hungry people to sign a treaty promising relief. And it would seem that both the Indians and the buffalo were seen as inferiors which could be disposed of as their superiors saw fit.

These differences in the way in which the land was valued were also connected to the way in which it was owned. Consistent with liberal individualism, the Western world owned the land privately; as individuals. The immediate separation of the land into private plots owned and controlled by individuals was everywhere the result of Western expansion into Native lands. This was not only consistent with Western individualism, but with its hierarchical pattern of relations as well. For it is through control of the land and its resources that control of other people follows. After all, the inevitable result of this pattern of ownership is that a significant percentage of the population is entirely dispossessed of any direct access to the resources of the land, and thus depend for their subsistence upon selling their labour to those who do own it. It creates a readily accessible labour pool among the poor for exploitation for profit or tribute by the land owning class.

Among First Nations such private ownership of the Earth itself was very rare–if not unknown. Even in agricultural societies such as those of the Pueblos, where families claimed access to plots and the products which they grew upon them, if the land went unused it became available for the use of others–it could not be sold. Consistent with egalitarian ideals, the land was considered to be a common inheritance which was accessible to all, so that even the poorest still had access to that which they required for subsistence. Indeed, the poor in most First Nations cultures were unable to acquire what they needed largely due to physical disability, old age, lack of skill, bad luck, or laziness, rather than because they did not have *access* to that which they required.

The subdivision of the land was often one of the motivations which lead Native Americans to resist the presence of the whites. It was an argument which Tecumseh put to good use, when trying to forge an intertribal alliance to resist the invasions of whites into the country of his people. As he stated, "The way, and the only way, to check and stop this evil, is for all the Redmen to unite in claiming a common and equal right in the land, as it was at first and should be yet; for it was never divided, but belonged to all for the use of each" (as cited in McLuhan 1971:85). This was not only consistent with the egalitarian ideals of Native American philosophies, but with the basic respect for the land and its inhabitants which was an integral part of their religions as well. Chief Joseph not only claimed that he had no right to sell the Earth because he did not own it, but felt that selling it was, once again, to play God:

> The country was made without lines of demarcation, and it is no mans business to divide it...The earth and myself are of one mind. The measure of the land and the measure of our bodies are the same...Do not misunderstand me, but understand me fully with reference to my affection for the land. I never said the land was mine to do with it as I chose. The one who has a right to dispose of it is the one who created it. I claim a right to live on my land, and accord you the privilege to live on yours (as cited in Brown 1970:316).

The land was not his to do with as he chose, for it belonged to the Great Spirit alone. Neither could it be rashly exploited to serve the ends of individuals, for in doing so one was exploiting the sacred power immanent in all living things, and in all of the creation. One thus had ethical obligations to the land itself, and not just to "persons," however narrowly or expansively that legal category might be defined. Indeed, it seems at times that Native Americans did not even conceive of the land as belonging to them in common, but rather that *they belonged to the land*. As Grand Chief Harold Turner of the Swampy Cree Tribal Council suggested in

1992, when speaking to the hearings of the Royal Commission on Aboriginal Peoples (RCAP) in The Pas, Manitoba:

> We as original caretakers, not owners of this great country now called Canada, never gave up our rights to govern ourselves and thus are sovereign nations. We, as sovereign nations and caretakers of Mother Earth, have a special relationship with the land. Our responsibilities to Mother Earth are the foundation of our spirituality, culture and traditions...Our ancestors did not sign a real estate deal, as you cannot give away something you do not own (RCAP:1997)[70].

The Native American attitudes discussed above were also consistent with a much more long term view. As Sitting Bull once stated, "I want you to take good care of my land and respect it...My children will grow up here, and I am looking ahead for their benefit, and for the benefit of my children's children, too; and even beyond that again" (as cited in Vanderwerth 1971:231-32). Or as Blackfoot of the Mountain Crow once opened a treaty conference, "The earth on which we walk, from which we come, and which we love as our mother–which we love as our country–we ask thee to see that we do that which is good for us and our children" (as cited in Vanderwerth 1971:194).

Once again we see the common Native American concern for balancing the interests of the group as a whole with those of individuals come to the fore. For the group is not only extended in space, but exists also over time. So its interests can only be long term, and when they are taken into consideration the interests of future generations must also be encompassed in one's deliberations. The Iroquois league, for example, had as one of its doctrines that any decisions that the council of tribes arrived at must be in the interests of the seventh generation of their descendants, and not only of the group as it was constituted in the

---

[70] This passage is from Vol. 2, Section 4, Chapter 3.1, "Lessons from the Hearings."

present (LaDuke 1999:198). And as Cajete adds:

> This 'seventh generation' principle of the great law of the Iroquois Confederacy is well understood by most Indigenous people as a prime directive of human and cultural sustainability. Cultures that violate this directive gradually cease to be and vanish over a period of time (2000:266).

Consistent with its emphasis upon the priority of the individual, however, the capitalist ethic of the Western world is short term. Since individuals live only a short time, decisions arrived at in terms of the interests of individuals need not even look ahead one generation. The decisions of Western politicians seldom extend beyond their current term of office, or beyond what is expedient for reelection when it expires, while those of economic planners often change quarterly. Consequently, as Arne Naess once observed:

> it is politically dangerous to be responsible for pollution that will clearly show itself within an election term, but it is much less politically dangerous to arrange things so that it will be the next generation or the generation after that who will suffer the real effects (1989:139).

Yet even 150 to 200 years ago, when Western technologies and lifestyles, and the destruction attending them were relatively benign compared to the present, Native Americans had already expressed their opposition to Western cultures' exploitive pattern of relations with the natural world. This illustrates not only a long term view, but also a very perceptive understanding of the relationship between humans and the natural world.

None expressed these sentiments more eloquently, perhaps, than Standing Bear. Keep in mind that these prophetic words were first published in 1930, almost two decades before the appearance of Aldo Leopold's early ecological classic, *A Sand County Almanac*, which some now refer to as the "bible" of the ecology movement. One should also remember that it is only since about the mid 1970s that the word "ecology" has entered into the common parlance of the Western world. The sensibility which informs Standing Bear's passage, on the other hand, is his interpretation of the cultural traditions of his people, which extend back into the mists of their collective memory:

> There was a great difference in the attitude taken by the Indian and the Caucasian toward nature, and this difference made of one a conservationist and of the other a non-conservationist of life. The Indian, as well as all other creatures that were given birth and grew, were sustained by the common mother–earth. He was therefore kin to all living things and he gave to all creatures equal rights with himself. Everything of earth was loved and reverenced. The philosophy of the Caucasian was, 'Things of the earth, earthly'–to be belittled and despised. Bestowing upon himself the position and title of a superior creature, others in the scheme were, in the natural order of things, of inferior position and title; and this attitude dominated his actions towards all things. The worth and the right to live were his, thus he heartlessly destroyed. Forests were mowed down, the buffalo exterminated, the beaver driven to extinction and his wonderfully constructed dams dynamited, allowing flood waters to wreak further havoc, and the very birds of the air silenced. Great grassy plains that sweetened the air have been upturned; springs, streams, and lakes that lived no longer ago than my boyhood have dried, and a whole people harassed to degradation and death. The white man has come to be the symbol of extinction for all things natural to this continent. Between him and the animal there is no rapport and they have learned to flee from his approach, for they cannot live on the same ground (1978:165-6).

III

THE CLOSING CIRCLE I:
ECO-HOLISM & NATIVE AMERICAN
PHILOSOPHIES

1. SHALLOW VERSUS DEEP ECOLOGY

> We stand at a crossroads in the evolution of Western consciousness. One fork
> retains all of the assumptions of the Industrial Revolution and would lead
> us to salvation through science and technology; in short, it holds that the
> very paradigm that got us into trouble can somehow get us out. Its pro-
> ponents...view an expanding economy, increased urbanization, and cultural
> homogeneity on a Western model as both good and inevitable. The other
> fork leads to a future that is as yet somewhat obscure. Its...goal is the
> preservation (or resuscitation) of such things as the natural environment,
> regional culture, archaic modes of thought, organic community structure,
> and highly decentralized political autonomy (Berman 1984:188-9).

THE WESTERN WORLD has increasingly come to realize that it is
faced with some serious ecological problems of late, there is as yet
little consensus as to the nature and extent of these problems, nor
concerning the extent of the changes required in order to effect a
relatively permanent and lasting solution to them. This chapter
shall attempt to illustrate that the changes necessary to reach an
adequate solution to ecological problems are much deeper and
more fundamental than is usually realized. This is because the
fundamental assumptions about the nature of reality upon which
an ecological world view are based represent a radical departure
from the traditional world view of Western culture. In fact, they
represent a readoption of views very similar to those of Native
Americans and other aboriginal peoples in many respects.

It is this very difference in the attitude which is highlighted by

Berman in the above quotation. His distinction is essentially the same as that made by Arne Naess between "deep ecology" and "shallow ecology." The distinction is between a type of reformist ecology–which feels that ecological problems are solvable within the current technological, economic, political and ideological framework of Western society–and a more revolutionary ecology which traces the root of contemporary ecological problems to this framework itself. It sees fundamental changes as necessary in *all* of these areas if we hope to reach an adequate solution to these problems. As Naess states it:

> Fundamental within the deep ecology movement is the insight that its goals cannot be reached without a deep change of present industrial soc- ieties...the aim of supporters of the deep ecology movement is not a slight reform of our present society but a *substantial reorientation of our whole civilization* (1989:151 and 45, original emphasis)[1].

In the shallow view, which he admits is currently dominant, the ecological crisis is seen merely as a "technical problem" which "does not presuppose changes in consciousness or economic system." Neither our ideology of consumption nor our practice of it are seen as problematic, and "continued economic growth is often taken for granted" (1989:96). This view is quite prevalent among politicians, business interests, in television advertising and the news media. It argues for the status quo: the economy will con-tinue to expand and "environmental" problems will be "fixed" by applying better engineering, better science, and better legislation. In the shallow view:

> the task is essentially one of 'social engineering,' modifying human be- havior through laws and regulations posed by ministries and departments of the environment–for the short-term benefit of humans (Naess 1989:162).

None of Western cultures' practices or aims require change in this view. For deep ecology, however, these beliefs and practices

[1] For his original discussion of the contrast see Naess (1973).

are seen as a fundamental part of the problem–indeed, the *cause* of the problem. Deep ecology attempts to suggest and promote a new world view and a new ethos, and to articulate their implications in practice, in the hopes of creating a lifestyle which is more adaptive to natural ecological conditions than that of contemporary industrial society. In doing so it questions the basic assumptions of our present ways of thinking and living.

For example, if Western technology has shown itself to be increasingly destructive of natural ecological processes, does it make sense to look upon more of the same as our savior? For the fact that the technology which Western culture has developed has proven itself to be the most ecologically devastating of any in the history of humankind is surely no coincidence. Rather, Western culture has developed the technology which it has because both its ideology and its patterns of social and economic organization tend to *promote* and to *value* such a development. Solving the "environmental" problem is not merely a technical problem–a question of the *means* we employ–but also a question of the *ends* which we pursue, and of the *manner* in which we pursue them as well. To borrow Hardin's (1968) reference to the population problem, the ecological crisis can be described as a "problem with no technical solution." It is not a question of the detrimental side-effects of any *particular* piece of technology, since Western technology has *repeatedly* proven itself to have detrimental ecological side-effects. The solution, therefore, will require an examination of the dominant *ideology*.

The essential problem from the perspective of deep ecology, then, is to understand the forms of sensibility which informed this development, the cultural values which the development pursues, and the patterns of organization through which it was realized. For in this view the ideological and the organizational contexts cannot be isolated from the technical, since changes in one sphere–such as technical changes in our relationship to nature–require and imply analogous changes in our patterns of social relationship, and in the world view and ethos which informs that behavior. Here again the ideological, the social and the ecological relationships are best pictured as a series of nested spheres–as circles within circles.

So we can expect no lasting solution to these problems from within the current status quo, for the simple reason that it is from the nature of the status quo itself that the problems arise. As Bookchin observed, "[t]he crises are too serious and the possibilities too sweeping to be resolved by customary modes of thought–the very sensibility that produced these crises in the first place" (1991:41). Further, a shallow view:

> does not bring into question the underlying notion of the present society, that humanity must dominate nature; rather, it seeks to *facilitate* that notion by developing techniques for diminishing the hazards caused by domination (1980:58-9, original emphasis).

The shallow view seeks not to *eliminate* the ultimate causes of the problems in the long term, but merely to *mitigate* detrimental effects in the short term. As should be evident, Naess' contrast between shallow ecology and deep ecology is not unique to deep ecology, but is one of the defining features of what I identify as an "eco-holist" position as well. Eco-holism does hide behind a variety of different names in the philosophical literature–including deep ecology (Devall 1980; Devall and Sessions 1985; Evernden 1993; Fox 1986, 1990; Naess 1973, 1989; Zimmerman 1994), social ecology (Bookchin 1980, 1986, 1990, 1991; Clark 1984, 1992) and eco-feminism (Beihl 1991; Diamond and Orenstein 1990; Griffin 1978; Merchant 1980, 1992). There is a also a growing literature in philosophy of science (Berman 1984; Capra 1975, 1982; Griffin 1988), biological ecology (Holling 1973; Holling and Sanderson 1996; Holling et al. 1998), and anthropology (Bateson 1972, 1977; Bateson and Bateson 1988; Ingold 1990, 1992; Harries-Jones 1992, 1995; Rappaport 1979, 1994) that is consistent with their position. All recognize this split within the larger "environmental" movement.

Indeed, the contrast between deep and shallow ecology is also precisely analogous to what Aldo Leopold originally dubbed "the A-B cleavage" among conservationists of his day in *A Sand County Almanac*, his early ecological classic (1949:221-23). Leopold distinguished between conservationists for whom the land was just "soil," and its function "commodity production," and another group for whom the land was viewed as something "much broader," or as an ecosystem with a wide variety of functions.

His original distinction has assumed a variety of different names among his followers. Johnathon Porritt and David Winner, distinguish between the "light greens" and the "dark greens" (1988:9-13). According to E. F. Schumacher the contrast is between the "forward stampede" and the "homecomers" (1973:155), while Murray Bookchin dubs the shallow view "environmentalism," while the word "ecology" is reserved for his own more radical social ecology or "ecological holism" (1980:57-71)[72]. Perhaps the most succinct and explicit contrast is provided by Wendell Berry–between the *exploiter* and the *nurturer*, which certainly requires little elaboration to get the point across (1977:7-8)[73].

In each case, the contrast is essentially the same, and this shared contrast serves to unite them all within a larger literature, which identifies the fundamental nature of ecological problems in a

[72] Bookchin also refers to his views as "ecological holism" in The Ecology of Freedom (1991:3).
[73] For a more detailed discussion of the many faces which this contrast has taken see Fox (1990:22-40).

consistent manner, and which shares many of the same objectives and aims–eco-holism. As should not be surprising, it is also the eco-holist point of view which has learned the most from Native American philosophies. For since reformist environmentalism does nothing to question the basic assumptions of Western industrial culture, it is unlikely to value Native American contributions which, as we have seen above, often call these assumptions into question. Eco-holism not only questions these assumptions, and the lifestyles consistent with them, but does so in a manner which echoes the criticisms which Native Americans have long made concerning Western ways of life.

Even the contrast between a human-centered view of the relationship between humanity and nature and a more respectful and egalitarian attitude can be found in Native American literature long before it surfaced in the writings of eco-holists. It is essentially the same contrast as that drawn by Standing Bear in order to contrast the views of the Western world with those of his own people (as cited at the end of Section II:4). He too drew a contrast between a view in which humans are considered to be a "superior creature" and all other beings to be "of inferior position and title," with a view in which humans were seen as being "kin to all living things," which enjoyed "equal rights with himself." Like the eco-holists who followed him some decades later, he also felt that "this difference made of one a conservationist and of the other a non-conservationist of life." In other words, it made of one a *nurturer* and of the other an *exploiter* of life.

Thus, it should not surprise one if the traditional philosophies of Native Americans and other aboriginal peoples have often served as models for the development of a more ecological sensibility in the eco-holist literature, while their lifestyles often serve as models for more adaptive patterns of relationship in the social and ecological spheres. Bookchin, for example, refers to the patterns of organization of aboriginal peoples as "organic societies," since he considers their egalitarian and participatory philosophies to be consistent with ecology[74]. As he states:

> their outlook, particularly as applied to their communities' relationship with the natural world, had a basic soundness—one that is particularly relevant for our times...I am eager to determine what can be recovered from that outlook and integrated into our own 1991:13-14)[75].

The similarities in outlook which eco-holist and Native American philosophies share, and their common differences from the dominant ideology of Western culture, are thus no coincidence, and this shall be a recurring theme throughout what follows.

---

[74] Though not in Emile Durkheim's sense of "organic solidarity," or a complex society made up of disparate parts.

[75] Bookchin makes no reference to Native American literature, however, to support his claims.

2. CIRCLES WITHIN CIRCLES: TOWARDS AN ECOLOGICAL WORLD VIEW

> Humanity has passed through a long history of one-sidedness...The great project of our time must be to open the other eye: to see all-sidedly and wholly, to heal and transcend the cleavage between humanity and nature that came with early wisdom (Bookchin 1991:41).

PERHAPS THE BEST WAY of illustrating the complementarity of ecological and Native American views, and contrasting them with that of Western culture is provided by Berman (1984), with his contrast between the *mechanistic* and the *organic* world views most typical of each. A world view consists of the fundamental assumptions about the nature of reality which a particular culture or philosophical outlook entertains, and it is precisely in this area that eco-holism begins to revise the traditional views of the Western world, and especially its understandings of science[76], in a direction which is more consistent with Native American views.

The birth of the mechanistic world view is usually traced to the advent of modern philosophy in Europe, and in particular to the works of René Descartes. Descartes is not only widely acknowledged as the "father" of modern philosophy, he is one of the earliest and most influential exponents of mechanistic views as well. Along with such contemporaries as Francis Bacon and Thomas Hobbes,

---

[76] Due to the manner in which it critiques the premises of modern science and suggests reasons for their modification, the eco-holist position is usefully viewed as building upon the work of the philosopher of science, Thomas Kuhn (1970). Kuhn argued that the history of science was not simply a matter of a linear progression, with one discovery merely building upon previous discoveries. Rather, Kuhn suggested that the actual history of scientific thought has been characterized by a series of "scientific revolutions," or "paradigm shifts," through which its understandings of the nature of the world, and of the activity of science itself, have been fundamentally altered. The classic example is the shift from Newtonian to Einsteinian understandings of physics. Eco-holism can be characterized as calling for a further paradigm shift from a mechanistic to an organic understanding, particularly of complex organic systems.

Descartes helped to articulate a new way of conceiving of the world.

Their views were quite unique even in the history of Western philosophy, and were closely associated with the rise of the new scientific outlook and methodology. The mechanistic views which they advocated, spurred on by Isaac Newton's spectacular success in applying them to the laws of physics, were to change the direction of Western civilization[77]. Present day science and technology owe their existence to the ideas of men such as these, who broke with previous traditions to forge a new world view, which transformed the earlier themes of separation and opposition into their most extreme forms, and cleared the way for these unprecedented developments.

To the modernists, nature came to be pictured as a machine, which moved according to mechanical laws. These laws could be understood, it was thought, and through such an understanding the forces of nature could be harnessed and manipulated to serve human ends. As Descartes put it, through the application of scientific methodology, humanity could become the "masters and possessors of nature." Or in the words of David Hume, "The only immediate utility of all sciences is to teach us how to control and regulate future events by their causes" (1962:89).

One should remember, too, that this philosophy did not develop in a vacuum. Rather, Descartes writings in the mid-seventeenth century, and their subsequent development and acceptance by European academia, were paralleled by the development of technology in the Industrial Revolution, the rise of capitalist organization, ownership and ideals, the expansion of

---

[77] Holling et al. (1998:344) similarly point out that modern science is based upon a combination of Descartes' philosophy, Bacon's experimental method and Newtonian physics.

Europe's colonial empires around the world, and the attendant domination of Native Americans and other Aboriginal peoples. It is in this context that mechanistic views originally gained acceptance and expanded their popularity, and it is in this context that their "usefulness" was originally demonstrated[78].

Following Berman, "I do not wish to suggest that Descartes is the lone architect of our current outlook, but only that modern definitions of reality can be identified with specific planks of his scientific platform" (1984:11). For Descartes' works not only set the program and problematique for much of subsequent modern philosophy, certain of his premises can also be identified in the works of many contemporary scientists and philosophers up to the present day.

Chief among these is atomism, which assumes that all things are essentially separate, as discussed above. This premise is more than evident in Descartes' definition of a substance, for as he states, "By *substance* we can understand nothing other than a thing which exists in such a way as to depend on no other thing for its existence" (1988:177). In order to qualify as a substance–in order to be considered *real*–a thing must be totally separable from its context, and able to exist independently and in isolation. Similarly, in order to properly understand things, one must first break down any larger wholes into their constituent parts, which can then be manipulated and reassembled to suit the experimenter's ends. For

---

[78] Interestingly, Nikiforuk has suggested that the Black Death, which recurred several times with decreasing intensity in Europe following its first occurrence in the early 1300s, may have helped to make Europeans sympathetic to mechanistic thought. As he puts it: "Retribution [for the plague's decimations] became the goal of Europe's new scientists. Descartes' philosophical view of the universe as a giant clock that could be set or reset to suit human ends had all the marks of post-plague thinking: if you don't like the time, just move the hands. After the plague, its not really surprising that humans decided to trust machines more than Nature" (1996).

it is the parts of which one can be most certain.

Consistent with this premise, Descartes began his inquiries not with the larger world around him, nor with the society of which he was also a part, but with his own individual mind, stripped of all attributes except self-awareness. The resulting separation of the human person into two substances–mind and body–provides a further illustration of his basic metaphysical premise as well–the basic *separateness* of all phenomena. According to Descartes:

> each substance has one principle property which constitutes its nature and essence, and to which all its other properties are reduced. Thus extension in length, breadth and depth constitutes the nature of corporeal substance; and thought constitutes the nature of thinking substance. Everything else which can be attributed to body presupposes extension, and is merely a mode of an extended thing; and similarly, whatever we find in the mind is simply one of the various modes of thinking (1988:177).

Hence the name "extended substance" is attributed to the body, while the essential attribute of mind is to think. Descartes not only sees the world as being essentially divisible into bits, but as divided into two distinct types of substance, or two distinct "realms of being" as well. This separation was derived from the premise: "I think, therefore I am." For Descartes found that even while he doubted everything about his body, this statement was still true each time that he conceived it or pronounced it in his mind. And since he believed this "I" must refer to something, he proposed that it referred to a "thinking substance;" a transcendent mind or soul. Thus, Descartes modeled his understanding of thinking sub-stance after Christian conceptions of the soul–as being totally unextended in space, as having no place, no mass and no matter, and as transcendent, or as not being part of the physical world at all. As he stated, from the above proposition, "I knew that I was a substance whose whole essence or nature is solely to think, and which does not require any place, or depend an any material thing in order to exist" (1988:36).

Yet it is not only the mind and the body which are separated and opposed by this contrast, but culture and nature as well, for such thinking substances are confined to human souls alone. All other things in nature are reduced to mere extension, with any qualitative differences between them being reduced to measurable spatial quantities. For it is extension, or measurability, which is the essential attribute of bodies. Similarly, larger unities are reduced to mere aggregates of parts, which impact upon one another like Hume's billiard balls to set the whole in motion. So as Holling et al. suggest, "the general conception of reality from the seventeenth century onward saw reality as a multitude of separate objects assembled into a huge machine" (1998:344).

While humanity escapes from this causal nexus through being gifted with a mind or reason, which is located "beyond" this realm of efficient cause, as a first cause in its own right, all other things are reduced to mere matter and motion. All of reality–from rocks, to trees, to animals, to the human body itself–is reduced to a machine. It becomes a mere aggregate of parts without meaning, purpose, direction or life. For only the transcendent human mind is considered to be the true subject of a life, and this divine spark belongs to humanity alone. The rest of the world, including even our own bodies, is seen as being essentially dead.

Hardly surprising, then, that eco-holism rejects the mechanistic view and the metaphysical presuppositions upon which it is based. After all, ecology is essentially an *organic* outlook, defining itself as the study of the interrelationships among *living* beings, and be-tween living beings and the inorganic environment. Indeed, what has fallen between the cracks in the mechanistic view, in which transcendence is seen as the only alternative to mechanism, is our life *in* the world, and any hope of adequately understanding it. For in the mechanistic world view, in a very real sense, only the *measurable* is real.

As G. W. Leibniz, an early critic of mechanistic views, was among the first to point out, the problem with this view is that the reduction of all (non-thinking) substance to mere extension strips it of all qualitative differences. As he described the implications of this view:

> under the assumption of the perfect uniformity of matter itself, one cannot in any way distinguish one place from another, or one bit of matter from another...no observer, not even an omniscient one, would detect the slightest indication of change" (1980:164).

Not only would spatial differences be masked by this uniformity, temporal ones would be as well. For without qualitative differences among things there would be no way of detecting any difference either between one thing and the next, nor between one moment and another. Indeed, if matter were truly completely uniform, consisting only of extension, there would also be no way in which it could be *measured!* Leibniz thus accuses the mechanists of reducing reality to "a night in which all cows are black," to use Hegel's phrase. For Leibniz, however, "such consequences are alien to the nature and order of things, and...*nowhere are there things perfectly similar*" (1980: 164, original emphasis).

In other words, qualitative differences are an essential attribute of all substance; of all reality. Indeed, the difference between the organic and the inorganic–between life and death–is just such a qualitative difference. After all, life is not an *amount* of *anything*. Nor is it an observable or measurable object. The difference between a living body and a corpse, for example, cannot be encompassed by quantitative criteria. As far as quantitative measures are concerned the corpse may be exactly the "same" as the living body. Its weight, height, volume and shape remain unchanged immediately after death. One of the only detectable quantifiable differences would be a drop in temperature in the corpse, but its *life* will forever elude quantification. Insofar as the purely quantitative sciences are concerned, then, the distinction between living and dead is a nonentity. So as Leibniz originally said of the mechanists:

> I am the most readily disposed person to do justice to the moderns, yet I find that they have carried reform too far, among other things, by confusing natural things with artificial things, because they have lacked sufficiently grand ideas of the majesty of nature (1980:141-2).

Appropriately, it is in the physical sciences–physics, chemistry and mechanical engineering–where the mechanistic methodology of modern science has had its greatest success in learning to predict, control and manipulate the phenomena it studies. Indeed, it is principally these sciences, along with applied mathematics, which are responsible for the spectacular developments in technology and engineering which the past few centuries have witnessed. As Schumacher observed, "the strength of science...derives from a reduction of reality to one or the other of its many aspects, primarily the reduction of quality to quantity" (1973:256).

While Schumacher would agree that the resulting blindness of modern science to such an important qualitative distinction as that between the living and the dead renders its methodology inappropriate for application to organic aspects of reality, it is important to note that he does not deny that the physical sciences can explain many things. This includes many things which apply to living organisms. If a person and a rock are thrown from a cliff, for example, they will both accelerate towards the ground according to the laws of physics–but the rock is unlikely to cry out in fear and flail its limbs on the way down. Yet *this* difference, being a qualitative one, the physical sciences have never been adequately able to explain.

Nor should one *expect* them to be able to do so, for they have an essentially mechanistic view of the phenomena which they investigate. *All* qualitative differences are reduced to quanta, despite that fact that life is unmeasurable. It clearly follows that this type of mechanistic epistemology, which knows the world only through quantification, can lead to an adequate and relatively complete understanding *only of those things which are dead*. In the mechanistic view, however, nature is conceived of as *essentially dead* in any case. In this way its epistemology, or its way of knowing the world, remains consistent with its ontology, or its assumptions about the nature of reality.

Since the phenomena which ecology is most interested in understanding are the organic interrelationships which give rise to and sustain living beings, however, the premises of the mechanistic view are obviously inadequate to its appointed task. For in the mechanistic view, extended substance is also seen as essentially passive. Like human artifacts, nature is inert, existing only to be manipulated and used for human benefit. It is something which is *acted upon*, not something which *acts*. The ability to act–to freely and spontaneously pursue ones own course–is reserved for the transcendent human mind alone. All other things are seen as governed by the inexorable laws of efficient causation, and have no natural spontaneity–no *life*–of their own. Any movement or change is to be explained only according to efficient cause, or by a force acting upon a thing from the *outside*, as in Hume's famous example of one billiard ball striking another.

Yet while efficient cause may be appropriate to explaining the action of billiard balls, machinery and other inorganic phenomena, efficient cause alone is totally inadequate when applied to the organic sphere. For example, if one kicks a rock efficient cause is perfectly able to explain why the rock is sent flying by the force which your kick imparts to it. If one kicks a dog this is not so. For in the latter case the dog is not only sent flying by your kick, *it also runs away!* Indeed, *it may even turn and fight!* And *what* it does is dependent not only upon the force of your kick, but upon the *character* of the dog in question.

Thus, a different type of causality is necessary to describe this second type of motion. It is not a motion which arises exclusively from an external force impacting upon a passive object. Instead, it arises *within* the object, as a spontaneous action on the part of a living being. It is, in other words, a *final cause;* an *action* which seeks an *end.* In the example of the dog given above the end is either to escape the effects of a second kick, or to bite the person who kicked it so that it will not, hopefully, be kicked again. The rock, on the other hand, seems indifferent to these possibilities.

The patterns of relationship in the organic sphere are thus *qualitatively distinct* from those in the inorganic sphere, while yet *contained within them,* and the causality necessary to adequately describe those patterns of relationship are similarly different in kind, and not only in degree[79]. As Bookchin relates:

> To the extent that mechanism became the prevalent paradigm of Renaissance and Enlightenment science, the notion of final cause became the gristmill on which science sharpened its scalpel of 'objectivity,' scientific 'disinterestedness,' and the total rejection of values in the scientific organon...Efficient cause, removed from the larger ethical matrix of Aristotelian causality, was now conceived as the sole description for natural phenomena in kinetic motion (1991:285 and 287).

According to the mechanistic view's "exaggerated rejection of

---

[79] McNeil (1976:7), for example, argues that mechanistic models of causality, or "simple cause-and-effect analysis," are inadequate for understanding the interactions between disease organisms and human populations, and by implication, other compex organic systems.

all organicism," there is absolutely no directiveness inherent in extended substance. It pursues no values, seeks no ends, and is absolutely bereft of any purpose or meaning. All of these attributes are confined to the transcendent human mind. And it was precisely because he conceived of extended substance as essentially determined that Descartes found it necessary that the mind be transcendent, for he did not wish to deny *human* freedom and spontaneity. Through his dualistic metaphysic he secured a place for such spontaneity and freedom of the will *outside of* natural, mechanistic processes, in the "realm" of the mind.

But there are two ways of approaching this problem. One is to maintain that all causality is efficient causality, and that human freedom can only be explained through separating and opposing mind and nature, as Descartes did. The other is to point out that efficient causation alone is obviously inadequate to explain everything which is empirically observable, and that it is ithe causal hypothesis which the mechanists advanced which requires modification.

It is, after all, *only an hypothesis*. And if we simply realize that there *is* spontaneity and life *in* nature (such as ourselves), and if we do not wish to have recourse to transcendent ghosts in order to explain this possibility, then it seems to follow that the inexorable, mechanistic causation which was *assumed* to govern everything which is observable is inadequate to the task of explaining everything which *is*, in fact, observable. As a result, this causal hypothesis appears to have been *falsified*, and in need of modification. As Bookchin observed regarding organic causation:

> [w]hile it may certainly be mechanical, causation is more meaningfully and significantly *developmental*. It should be seen more as...an emerging process of self-realization, than as a series of physical displacements (1991:283, original emphasis).

Organic philosophy, then, finds it necessary to restore the idea
of final cause to nature; a notion which had, after all, always been
prevalent in ancient and medieval European philosophy prior to
the mechanists. As the ancient Greeks realized, natural organic
developments such as that from an acorn to an oak tree, or from a
human infant to a mature adult, cannot be adequately understood
purely in terms of efficient causality. They are not merely a matter
of the imposition of external forces upon essentially passive substances. Rather, they are guided largely by *internal* forces, or by an
internal *logic*. To make use of Aristotle's terminology, organic
development is *teleological;* it is not only impelled from behind, but
seeks a certain *end*. It is this activeness and directiveness which
distinguishes the inorganic from the organic–in which the
qualitative distinction between them lies. For as Janet Beihl relates,
even for such a simple organism as an amoeba:

> the very fact that it is maintaining itself is a germinal form of selfhood
> and a nascent form of subjectivity. Whatever else an amoeba may have,
> this active self-identity distinguishes it from the nonliving environment in
> which it is immersed (1991:117).

Thus, spontaneous activity, subjectivity, selfhood, mind and
eventually even freedom as humans know it are not *opposed* to
nature by eco-holists. Instead they are seen as *emerging from* it. In
such a view they are in no way separable from the larger processes
of organic evolution and development from which they originally
arose. As already noted, even the mechanists did not wish to deny
such spontaneity, if only among humans. That the founders of
science itself were willing to posit *ghosts* in order to explain this
possibility, however, is certainly a dramatic illustration of the
importance which they accorded to the hypothesis that all things
were essentially governed by efficient cause. For this hypothesis

left no room for such freedom in nature, and in order to remain free the mind had to be conceived of as a transcendent entity located "beyond" nature—as a "ghost in the machine" (Ryle 1949).

In an organic view, however, the human mind, like the human organism, is considered to be a part of nature. Thus, anything which can be attributed to the human mind must also be attributable to nature itself, and there is room for final cause, for *telos*, for immanent action and directiveness in biological organisms themselves, without need of transcendence. Similarly, if one is to speak in terms of "spirit," rather than "teleology," it must also be conceived of as immanent, or as a part of nature, as is common in Native American philosophies.

So in contrast to the atomism and dualism of the mechanistic view, the basic ontological assumptions of eco-holism are *relational* and *contextual*. It is, in a word, an *holistic* point of view. For as Bookchin notes, if one wishes to adequately understand *organic* phenomena, or complex organic systems:

> We can no longer afford to remain captives to the tendency of the more traditional sciences to dissect phenomena and examine their fragments. We must combine them, relate them, and see them in their totality as well as their specificity (1991:21).

In considering individual organisms to be parts within a larger whole, organicism does not take the opposite route to the mechanists, and reduce the parts to an undifferentiated "oneness." Instead, it seeks the middle course between the two extremes. As John Clark observes, "an organic view sees nature as an organic unity-in-diversity in a process of self-development" (1984:217). Neither parts nor wholes are considered to be basic, rather, they *imply* one another, for they are complementary terms. The whole could not exist unless it were made up of parts, and the parts could not exist apart from the larger whole. It is neither parts nor wholes alone which are the object of study, but *the relationships between them*. So an ecological view hardly ignores individual organisms, it merely focuses our attention upon *the nature of their inter-dependence*. Such patterns of interrelationship between organisms does nothing to erase their individuality. For this would be to ignore the many qualitative differences by which the parts are naturally distinguishable from one another, and to separate a consideration of the whole from a consideration of the parts.

As has often been stated, the fundamental ontological assumption of an ecological point of view is that *all things are connected*. This implies not only a contextualized and relational view, but also a participatory view in which the relationships within which each organism participates are considered to be essential aspects of what it is in itself, rather than as "external" to its nature. As Naess observes, an ecological view "is concerned first of all with relationships between entities as an essential component of what these entities are in themselves" (1989:36) Or again:

> Relationalism...makes it easy to undermine the belief in organisms or persons as something which can be isolated from their milieu...*an organism is interaction*. Organism and milieu are not two things–if a mouse were lifted into absolute vacuum, it would no longer be a mouse. Organism presupposes milieu (1989:56, original emphasis).

In other words, the relationship between an individual organism and the larger ecological context of which it is a part is an *enabling condition* for the life of that organism, and it is this relation-ship in which its *life* consists. One need look no further than one's life in order to confirm this. For with each breath of air which we take, with each glass of water we drink, and with each bite of food we eat, the dependence of our individual lives upon their ecological context is amply demonstrated. Each of these relationships demon-strates that our lives are intimately intertwined with the natural cycles in which we participate. The oxygen which we and other animals breathe is provided by plants, which produce the oxygen from the carbon dioxide which we and other animals have exhaled. The animal and the vegetable are thus complementary poles of a causal cycle, each of which produces what the other requires. Similarly, each time that we drink we participate within the larger hydrological cycle, within which all of the waters of the Earth are interconnected in a natural cycle of evaporation, transpiration, cloud formation, precipitation, run-off and evaporation. Inasmuch as the waters of the Earth quite literally flow through us daily, this interrelationship must also be considered to be an essential aspect of our lives, as it is of all life on Earth.

Through eating we are also intimately connected to other forms of organic life, consuming their substance so that we might continue to live ourselves. We are thus dependent not only upon those species which we directly consume, but also upon all of the natural conditions which *they* require for their existence. Finally, so as not to ignore the "dirty" half of these relationships; we also defecate, urinate, spit, sweat, bleed, clip our fingernails, blow our noses, lose our hair, shed our skin in tiny fragments, and in various other ways return our bodily substance to the Earth on a daily basis, even long prior to our death.

Thus, from an eco-holist point of view, our *life* does not end at our skin. Rather, it consists of these patterns of relationship with our organic and inorganic environments. Further, these relationships are no longer seen as merely contingent to our beings. They are instead understood to be essential aspects of our life itself. Quite simply, this is why we die if deprived of food, water or air for sufficient periods of time—for we are *internally related* to these things, and in these do our lives consist. Just as Holmes Rolston has observed:

> Savannas and forests are as necessary to elephants as hearts and livers. In the complete picture, the outside is as vital as the inside...The skin-in processes could never have evolved, nor can they remain what they are, apart from the skin-out processes. Elephants are *what* they are because they are *where* they are (1988:164, original emphasis).

Eco-holism therefore returns to a view which many Native Americans have long adhered to, for as the Lakota elder, Okute, put it in 1911:

> An animal depends a great deal on the natural conditions around it. If the buffalo were here today, I think they would be different from the buffalo of the old days because all the natural conditions have changed. They would not find the same food, nor the same surroundings. We see the change in our ponies (as cited in McLuhan 1971:19).

It appears that Okute was well aware of the fact that an important part of what an organism is, is not contained within its skin, but rather in its relationship to its ecological surroundings. And as Naess reiterates these points:

> There is no completely isolatable I, no isolatable social unit. To distance oneself from nature and the 'natural' is to distance oneself from a part of that which the 'I' is built up of...We are not outside the rest of nature and therefore cannot do with it as we please without changing ourselves (1989:164-5).

The "self" is considered to be *a part of*, rather than *apart from* nature. It is a circle within the larger circle of nature; a part within a larger whole. Indeed, what applies to individual humans also applies to human culture, for just as the organic sphere emerged from within the inorganic, so also human culture—or "second nature"—emerged from within the sphere of "first nature." The natural sphere is thus an enabling condition of the cultural sphere, with the relationship between them being that of part to whole, or of a smaller sphere contained within a larger. So as Rachel Carson once put it, "Man, however much he may like to pretend the contrary, is a part of nature" (1962:188).

From an eco-holist point of viewthe best illustration of the fundamental structure of reality, and of the manner of our relationship to it, is to picture it as a series of "circles within circles within circles, with no beginning and no end," just as Lame Deer has suggested. This symbolism is not only appropriate as an illustration of the way in which the self, and consequently human culture, are a part of nature–a smaller sphere contained within a larger–but is also the best description of the patterns of relationship in an ecosystem. For in a very real sense, the patterns of interrelationship in ecosystems do have no beginning and no end, but turn back upon themselves to form a circle. As Carson observes of the interrelationships which exist within the soil, and by extension, in the rest of the ecosystem; "[t]he soil exists in a state of constant change, taking part in cycles that have no beginning and no end" (1962:53). And as Lame Deer continues:

> This symbol, a circle within a circle, stands for life, for that which has no end...To a white man symbols are just that: pleasant things to speculate about, to toy with in your mind. To us they are much, much more. Life is a symbol to be lived...we are one with the universe, with all the living things, a link in the circle which has no end (1972:167 and 106-7).

Or in the words of Roy A. Rappaport, a cultural ecologist:

> the circularity of...ecosystemic structure blurs the distinction between cause and effect, or rather suggests to us that simple linear notions of causality...are inadequate...Among those being affected in unforeseen and possibly unpleasant ways may be the actor himself (1979:170).

In contrast, the mechanistic view always considered causality to be distinctly linear, as evident in the example of the billiard balls. It does not seem to consider the *game* of billiards itself, nor the manner in which that game is a rule-guided human relationship, in which the way that the balls are struck will have an effect upon one's next shot. For the mechanistic model has always been most concerned with how humanity can manipulate and control nature. Turning to social archetypes, rather than to natural ones, its model for all relationships is human artifacts and machines, and the manner in which they may be manipulated and controlled to serve human ends.

The effects which these interventions set in motion, however, have seldom been given the attention which they deserve. So the seeming linearity of their model is a result of the fact that they tend to ignore the latter half of this causal cycle, or at least to consider the effects of humans upon nature separately from the effects of nature upon humanity. Indeed, it was only with the advent of ecological thinking in the past few decades that the Western world has begun to give any serious consideration to the effects of its technology upon those who created it. As Fools Crow relates, however, in the view of his Lakota people:

> *Wakan Tanka* made the earth so that it does everything to take care of itself. Everything fulfills its role–ants, worms, vultures, wolves, pebbles, sand...when we damage any of these or the earth itself, we damage ourselves. We cannot use them up or waste and destroy them without paying a terrible price for it. But people are doing this, and it is coming back on us all. Grandmother Earth is crying out about it (1991:68).

But then the type of behavior which he describes is precisely what one would expect from people who conceive of both themselves, and their culture, as *opposed* to and *separate* from nature. Then one need not worry about the chains of determination one's actions upon the world set in motion affecting oneself–since they can never reach us. Similarly, if one conceives of the world as dead in any case, then no constraints exist on its exploitation, for one can do it no harm. It is for these reasons that eco-holists consider mechanism to be the dominant world view of Western culture. For whether or not we *explicitly* adhere to these views, as a culture, we *act* as though we believe them.

In an organic view, on the other hand, atomism is replaced with holism, dualism with participation, separation with relationalism, opposition with complementarity, and linearity with cyclicity (or recursiveness). These changes are made necessary not only because eco-holism has come to realize that everything *is* interconnected (as Native American philosophies articulated long ago), but from the fundamental fact that nature itself is conceived of as *alive*. So as McGaa describes the emergence of a more ecological sensibility in the Western world, from the perspective of a Lakota scholar following Black Elk's tradition and teachings:

> Brave new thinkers and independent spirits out in the non-Indian world were reading the tale of environmental destruction that was in progress...New values divergent from past values, new respects backed up by nonmaterial endeavors were the trademark of these early environmental pioneers–trademarks reflected as respect and care for nature. Humankind was not dominant over nature. Men and women were dependent upon nature, after all, and for some there grew the beginning of the belief held by the Indians–nature was a living entity (1990:129-30).

Indeed, the most fundamental similarity between eco-holism and the views of many Native American philosophies appears to be their shared organicism. As McGaa states, "You say *ecology*. We think the words *Mother Earth* have a deeper meaning" (1990:203, original emphasis). Both expressions do seem to imply similar attitudes and outlooks, however, for they share a fundamental assumption. This is that the world is not a mechanical phenomena, but an organic one–that everything is not fundamentally dead, but fundamentally alive. In fact, even such "inanimates" as air, water, sunlight and soil are in at least a special sense *alive* from an ecoholist point of view, since they are a necessary part of the complex organic systems of which life consists. For when life itself is conceived of as an empirically observable pattern of relationship, rather than as a metaphysical property confined within the skins of individual organisms, even the distinction between organic and inorganic begins to blur. This is because *everything* within the biosphere, or Mother Earth, participates within the patterns of interrelationship in which life consists.

Indeed, as J. E. Lovelock (1979) has observed, the biosphere itself can be considered to be a living organism, which actively maintains environmental conditions favourable to the perpetuation of its own life. And it is no coincidence that he names the living biosphere "Gaia," after the ancient Greek Earth Goddess. For Lovelock also seems well aware of the fact that the idea of a living biosphere, and the idea of Mother Earth, are essentially the same concept expressed in different terms.

So whether we speak of "ecology," "the biosphere," and "complex organic systems," and consider everything to be *interrelated*; or whether we conceive of "Mother Earth" and her children as "kin," and all living things as our *relatives*, we are essentially speaking the same language. As McGaa says of the philosophical traditions of his people, in words which are equally applicable to an eco-holist position, "[y]ou may call this thought by whatever fancy words you wish–psychology, theology, sociology, or philosophy– but you must think of Mother Earth as a living being" (1990:208).

3. AN ECO-HOLIST CRITIQUE OF INDIVIDUALISTIC AND ANTHROPOCENTRIC ETHICS

> The fact that man can have the idea 'I' raises him infinitely above all other beings living on earth. By this he is a person...that is, a being altogether different in rank and dignity from things, such as irrational animals, which we can dispose of as we please. So far as animals are concerned, we have no direct duties. Animals are not self-conscious and are there merely as a means to an end. That end is man...Our duties towards animals are merely indirect duties towards humanity (Immanuel Kant, as cited in Rolston, 1988:63).

> A thing is right when it tends to preserve the integrity, stability, and beauty of the biotic community. It is wrong when it tends otherwise (Leopold 1949:225)[80].

THE ABOVE QUOTATIONS represent, in nut shell, the ethical positions with respect to nature which are consistent with the two types of world view described above. The first is a classic statement of the traditional individualistic ethics of the modern Western world which are–in a word–*human-centered*. The second represents Leopold's famous statement of the *land ethic*, which has informed the development of eco-holism. It is the human-centered ethic which is currently dominant in Western society, and as Miller relates, "[s]uch an ethic views the universe as morally divided between the human circle, within which all ultimate moral ends are to be sought, and the rest of the world, whose primary importance is to serve those ends" (1983:319).

This view is perfectly consistent with the mechanistic view described above–humans are not only seen as *separate* from nature, but as *superior* to it as well. It is an essentially hierarchical view. Just as God is superior to his creation, or a king is superior to his subjects, humans are considered to be superior to the rest of nature. Indeed, the sole value which is attributable to nature outside of the

---

[80]. For additional discussions of the land ethic see Callicott (1989) and Hefernden (1982).

human sphere is contained in its usefulness to humanity; or its ability to serve the interests of its superiors. All value is humanly conferred, or relative to human ends, while nature itself is considered to be a moral desert, which is not only dead, but as barren of all values as well. Consequently, as Miller continues, "We have, in the West, no sacred places, apart from monuments to human deeds and aspirations or to a human-loving God, and we have no sacred cows" (1983:319).

Because the source of all values–the human mind–is considered to be separable from nature, humans may do with the natural world as they see fit, without bothering themselves about ethical concerns. Human interests may be freely maximized, while those of all other living things, and of the land itself, may be minimized and dismissed. For in this view, as William Baxter so succinctly put it, "Damage to penguins, or sugar pines, or geological marvels is, without more, *simply irrelevant*...Penguins are important because people enjoy seeing them walk about rocks" (1986:215). All values, therefore, are restricted to the human sphere, or to the sphere of *mind*. They are considered to be personal and subjective matters which, like the mind itself, are seen as being "outside" of the natural world. The "objective" world is seen as containing only "facts," while all values are restricted to the sphere of mind.

It is from this bifurcation separating spirit and material, or values and facts, that the persistent myth that science is a "value-free" or "value-neutral" enterprise was derived. According to this dualistic metaphysic, nature contains only brute facts, which it is the business of science to discover. As a consequence, as Berman relates, "[s]cience cannot, supposedly, tell us what the good life is. Of course, this modesty is highly suspect: 'value-free' is itself a value judgment, amorality a certain species of morality" (1981:255).

Indeed, consistent with the human-centered ethic, its aim becomes the manipulation and control of nature to serve human ends. Yet questions concerning *which* ends we ought to pursue and *why* we ought to value and pursue these ends rather than some others are rendered taboo, since they become merely "subjective value judgments" rather than "objective matters of fact." For in the scientific enterprise it is the "objective" which is to be maximized, while the "subjective" is to be minimized or dismissed. Ethical concerns are considered to lie outside of the province of science, which becomes almost exclusively preoccupied with technique. As Bookchin points out, in the view of the mechanists, reason itself has become:

> merely a technique for advancing our personal opinions and interests. It is an instrument to efficiently *achieve* our individual ends, not to *define* them in the broader light of ethics and the social good...Not only do means become ends, but the ends themselves are reduced to machines (1991:270, original emphasis).

The manipulation, control and domination of nature have become ends-in-themselves, and science, shorn of all overt values, has dedicated itself to the development of ever better techniques and technologies to facilitate that domination. Whether the domination of nature is an end which we ought to pursue is a question which traditional science has consistently failed to entertain–as in the case of the scientists who developed the atomic bomb, for instance–for questions of fact must be kept strictly separate from those concerning values. Thus "why" is replace by "how" as the central scientific question, and ethical questions are replaced by an instrumental logic which emphasizes technique, manipulation and management–means separated from the ends they serve. As Vine Deloria Jr. describes the white man's science: "[h]is science creates gimmicks for his use. Little effort is made to relate the gimmicks to the nature of life or to see them in historical context" (1970:189).

Where the questions of traditional science end, however, is the point at which those of eco-holism begin, for from an eco-holist point of view questions of fact are considered to be fundamentally *related* to questions of value in the first place. As Beihl observes, from an eco-holist perspective "nature reenters the philosophical and political sphere of Western culture as an ontological ground for ethics" (1991:125). In other words, one's ethos is informed by, and intimately related to, one's world view. So if one's conceptions of the nature of the world change, then one would expect that this would imply analogous changes in the manner in which one treats it as well–particularly if one sees it as an organism rather than as a machine. And if all things are seen as fundamentally related to one another, rather than as fundamentally separate, this also implies that facts are related to values. The task is then to describe the nature of this relationship, and as Naess suggests, "it becomes important in the philosophy of environmentalism to move from ethics to ontology and back" (1989:67).

This is an important example of the application of holistic or relational thinking–the hallmark of an ecological approach–to philosophical concerns. For if all aspects of nature are systematic-ally related, then ontology and ethics–facts and values–cannot be considered in isolation from one another. Neither can science re-main isolated from ethical and philosophical concerns, for as Naess continues; "At the end of the scientific process lie ultimate assumptions of a philosophical kind" (1989:40). As the previous two sections have attempted to illustrate, its ultimate assumptions are those of the mechanistic view, with all of its attendant dualities in tow. From an eco-holist perspective, on the other hand, all of the traditional dualities in the Western philosophical tradition are systematically dissolved, with eco-holism generally seeking a balanced understanding of the relationships between these various "separate" areas. Its understanding of such pairs-or-opposites, is thus one of complementarity and relation, rather than of opposition and separation, since it employs a relational symbolism similar to that of the Native America. As Clark relates:

> In affirming such a holistic approach, Social Ecology rejects the dualism that has plagued Western civilization from its beginnings: a dualism that sets spirit against matter, soul against body, humanity against nature, subjectivity against objectivity, and reason against feelings; a dualism that is intimately related to the social divisions that are so central to the history of civilization—ruler versus ruled, rich versus poor, urban versus rural, civilized versus savage, male versus female; in short, the dominant versus the dominated (1992:91).

An eco-holist perspective not only questions the dualistic metaphysic which underlies the mechanistic world view, but rejects the logic of domination with which it is consistent as well. It is not a hierarchical view but an egalitarian one. For in a relational view humanity can no longer be considered to be either separate from nor superior to nature, nor can its interests be maximized at nature's expense. Rather, as Leopold himself once stated, "a land ethic changes the role of *Homo sapiens* from conqueror of the land-community to plain member and citizen of it. It implies respect for his fellow-members, and also respect for the community as such" (1949:204).

The land ethic is eco-holism's reply to the traditional human-centered ethics of Western philosophy. Having already adopted the organic, participatory and relational views of many Native American thinkers, eco-holism also finds itself adopting an ethic which is broadly similar to the one which has always remained so characteristic of traditional Native American philosophies—a direct respect for the land itself. Indeed, as Standing Bear relates, "Every true student, every lover of nature has 'the Indian point of view,' but there are few such students, for few white men approach nature in the Indian manner" (1978:195-6). Eco-holists appear to number among those few, for "[e]thics is not complete until extended to the land" (Rolston 1988:188).

Perhaps the best way to begin a description of the land ethic as developed by eco-holists is to first consider some of the things which it is *not*. For there are many points of view within the broader field of "environmental ethics" which do not adhere to the land ethic, yet which do attempt to extend ethics beyond the human sphere. Most represent little more than a continuation of the traditional ethics of the Western world, and do little or nothing to question the basic assumptions of the Western ethos, so the distinction between shallow and deep ecology remains relevant throughout what follows.

J. Baird Callicott distinguishes between three approaches within the broader field of environmental ethics; these being the traditional human-centered approach, the animal rights movement and the adherents of the land ethic (1989:15-38). The first group, as the name suggests, does nothing to question the traditional assumptions of Western ethics; it simply applies them as is to environmental problems. Baxter is an excellent example of such an approach, for as he states, "I reject the proposition that we *ought* to respect the balance of nature or to preserve the environment un-less the reason for doing so, express or implied, is the benefit of man" (1986:216, original emphasis). Thus, his position remains perfectly consistent with the doctrine of human superiority, for the environment's only value rests in its value to humans, and it is for this reason alone that it should be preserved. Yet it need not be preserved intact, for only those portions of nature which are valued by humans are of any concern. According to Baxter, "our objective is not pure air or water but rather some optimal state of pollution" (1986:216). His argument applies a typical cost-benefit analysis to the problem of pollution, which weighs the satisfaction humans gain from having a cleaner environment against the satisfactions lost by giving up certain material goods. As Baxter states it:

> trade-off by trade-off, we should divert our productive capacities from the production of existing goods and services to the production of a cleaner, quieter, more pastoral nation up to–and no further than–the point at which we value more highly the next washing machine or hospital that we would have to do without than we value the next unit of environmental improvement that the diverted resources would create (1986:218).

What goes unmentioned is that according to his own criteria one could successfully argue that we have such an "optimal" level of pollution *right now*. For it is *precisely* our tendency as a society to prefer the next washing machine and hospital to the protection of the environment which has led to the ecological crisis in the first place (or perhaps more aptly, the next pail of pesticides, the next nuclear reactor or missile, the second car, the third television, and the disposable diapers...). As a consequence, "trade-off by trade-off," the environment is being destroyed through pollution and the overconsumption of resources.

To the extent that Baxter's cost-benefit analysis is based upon an economic model, he essentially argues that we should "let the market decide," and if humans value their material goods more highly than rain forests, then the rain forests be damned! Thus Baxter does absolutely nothing to question the assumptions of the ethical status quo, the materialism of the Western world, nor the sensibilities which led to the problems of pollution and over-consumption in the first place, and does more to obscure the nature of ecological problems than to solve them.

In contrast, the animal liberation movement is at least a step in the right direction, though it also falls far short of being a truly ecological ethic. In fact, it does not appear to be a response to the ecological crisis at all. As Callicott observes, though animal rights views do challenge the notion of human superiority, and attempt to *expand* the sphere of ethical concern by extending certain rights to sentient animals, "there is no serious challenge to cherished first principles" (1989:20). Instead, the animal rights movement appears to represent an *extension* of political liberalism, which inherits the individualistic biases of its predecessors (and individualism is simply atomism applied to social thought, where it is the individual which is of primary concern, rather than the larger society or community). Thus, animal rights does nothing to challenge atomistic premises, but argues that *given* such premises it logically follows that moral consideration must be extended to include (at least some) animals.

Two of the best known exponents of animal rights views are Peter Singer and Tom Regan. Singer's position is an extension of the utilitarian ethics of Bentham and Mill, which were considered above. Their ethics were hedonistic, considering the ultimate moral end to be the maximization of human pleasure. Further, pleasure was considered to be always and everywhere good, while pain was considered to be the very substance of evil. Pleasure, therefore, was to be maximized and pain minimized. Since many animals also feel pleasure and pain, Singer reasons that all sentient creatures should be extended moral consideration, for the pain of laboratory animals is no less real and no less painful than our own. As a consequence, their pain must also be taken into consideration when calculating the net utility or disutility of one's acts. Considering the fact that, as Singer observes, even Bentham himself had speculated upon this eventuality in his original formulation of utilitarianism, the inclusion of sentient animals in calculations of general utility does seem consistent with his premises (1986:24-32). Such a position would, however, allow for tradeoffs in the usual economistic manner of the utilitarian cost-benefit analysis.

Concerned with these tradeoffs, perhaps, Regan goes further, offering a Kantian style rights ethic as a foundation for animal rights, and encouraging us to extend Kant's *categorical imperative* to animals as well as humans. Like Singer, Regan accuses traditional Western ethics of "speciesism," which is defined as an arbitrary and unjustifiable preference for one's own species–just as racism is an unjustifiable preference for one's own "race," or sexism an unjustifiable preference for one's own sex. This criticism would apply to Kant himself, as should be evident from the introductory quotation to this chapter. Regan thus rejects personhood, or mere membership in the human species, as a criterion for moral standing. His criterion is that the animal be "the experiencing subject of a life," or a nascent ego. All conscious animals are thus encompassed by Kant's categorical imperative, and must, in all instances, *be treated as ends-in-themselves, and never merely as a means* to our own selfish purposes, as Kant originally enjoined for humans. Regan therefore encourages the immediate and total abolition of commercial stock raising, hunting and trapping, as well as the use of animals in scientific experimentation (1986:32-9 and 203-5).

Once again, however, the primary focus of ethical concern remains sharply focused upon individuals. Nor can their ethics be expanded to encompass ecological concerns. Rather, such positions are essentially responses to human mistreatment of individual animals, and are concerned with the pain which we inflict upon them. Similarly, though they do challenge "speciesism" or human-centeredness, they do nothing to challenge the hierarchical logic of traditional Western ethics. For as Regan states, in the animal rights movement, "it is our similarities, not our differences, that matter most...we are each of us the experiencing subject of a life...the same is true of those animals which concern us" (1986:37).

For Kant, the fact that humans could say the word "I," or the fact that we are self-aware, was the essential shared attribute of all humanity from which moral considerability was believed to arise. For Regan, on the other hand, it is not self-awareness but sentience or consciousness which is important, which is an attribute which humanity shares with other animals. Yet they may only be considered worthy of moral regard–as equals–on the basis of their *similarity* to humans, for to be *different* continues to imply their *inferiority*. This is a point which is made abundantly clear in the following quotation from Singer's works:

> A life with no conscious experiences at all...is a complete blank; I would not in the least regret the shortening of this subjectively barren form of existence...The life of a being that has no conscious experiences is of no intrinsic value (As cited in Rolston, 1988:94).

Consequently, all beings who are capable of feeling pain may be equal, but all those who cannot remain inferior and irrelevant–mere resources for the use of this superior class of creatures. As usual, the interests of those within the sphere or moral con-siderability are to be maximized at the expense of those without it. For it is not *life* but *consciousness* which is respected, and while sentient animals are considered worthy of moral regard, ethical concern ends there. So while animal rights views reject a prejudice in favour of one's own species as arbitrary, they replace it with an equally arbitrary prejudice in favour of beings with a nervous system sufficiently developed for them to be capable of feeling pain. As Rolston suggests, therefore, "[w]e give animals rights only so far as we can see ourselves in them" (1988:62).

Besides the fact that animal rights views do nothing to answer ecological concerns, however, Callicott raises a further criticism as well. This is his rejection of the assumption that pain is always and everywhere evil, and that the world would be a better place if it were completely eliminated. For as he states, this hypothesis "is biologically preposterous. A living mammal which experienced no pain would be one which had a lethal dysfunction of the nervous system" (Callicott 1989:32). Pain is thus accepted as a natural fact of life. It is not the substance of evil, but a biologically important process. In fact, it is a natural warning system which informs us when our body has been damaged, and when the damage has healed, and in the larger scheme of things we would actually be much worse off without it. As Callicott continues, "[i]f nature as a whole is good, then pain and death are also good," for they are a necessary part of life (1989:33). Yet they are, as Rolston puts it, a "sad good." This view does not imply, therefore, that we should inflict pain and death upon animals or upon each other without any moral concern, for all things must be respected. It is simply intended to illustrate the fact that holding up the complete elimination of pain and death as an ethical ideal is entirely unrealistic. The lamb will never lie down with the lion, and it is ecologically preposterous to hope that it ever would.

Similarly, if humans eat other animals they are only doing what hundreds of other species in nature do, for being carnivorous is just as natural as being a herbivore. After all, the Earth has evolved many species which survive by eating the flesh of other animals. Indeed, many of the animals with the most highly developed nervous systems are themselves carnivorous. Yet wolves, lions and other carnivores would apparently have to be summarily condemned by the animal rights movement as incorrigible murderers, and be exhorted to take up a vegetarian diet along with the rest of us. Of course, such consequences are clearly ludicrous from an ecological perspective, in that they "imply the ecological nightmare of a policy of predator extermination" (Callicott 1989:277). Yet as Callicott observes, they do reveal that animal rights views "betray a world-denying or life-loathing philosophy. The natural world as actually constituted is one in which one being lives at the expense of others" (1989:33). Nor could it be otherwise, for vegetarians must also kill in order to survive.

Thus, for Callicott and other followers of the land ethic, the animal rights movement is not considered to be an ecological position at all, but merely a continuation of the liberal ethics typical of the modern Western tradition. It is not an attempt to *change* traditional ethics in any fundamental way, but rather an attempt to *extend* it to the "higher" animals. Its debate with more traditional humanism is thus an *internal* debate, which has served to obscure "the much deeper challenge to 'business-as-usual' ethical philosophy represented by Leopold and his exponents, and to keep ethical philosophy firmly anchored to familiar modern paradigms" (Callicott 1989:36).

Perhaps the most useful lesson which can be learned from an animal rights position is from the manner in which it fails. For it cannot encompass ecological concerns precisely because it retains the atomistic assumptions of the human-centered position. And this amply illustrates the fact that traditional liberal ethics, even when extended as far into nature as they can logically be extended, are inadequate to the task of understanding the change in values which the ecological crisis has made necessary.

A further expansion of the ethical sphere is attempted by Paul Taylor (1986). Unlike the animal rights movement, Taylor wants to expand the sphere of ethical concern to encompass not only individual animals, but all living organisms. His position is again another step in the right direction, and thus exhibits many more similarities with an eco-holist position. Principle among these is the proposition that life itself has intrinsic value, with which eco-holism is in perfect agreement. Both positions are, in a sense, biocentric or life-centered positions, which consider not self-awareness and personhood, nor sentience to be the basis of value and moral considerability, but rather life itself. The question for Taylor thus becomes not "Can they suffer?;" but rather "Are they alive?"

Yet his conception of what life is, and therefore of which aspects of nature are considered to be worthy of respect, differs sharply from that of the land ethic. An explanation of these two different conceptions of life and its value will allow for more direct comparisons with the land ethic, however, since the explanation of their differences will necessarily involve an explanation of some of the fundamental values of the eco-holist position as well.

Though Taylor claims to adopt several of the key premises of an eco-holist view, he does not seem to consistently follow out their implications. In particular, his "biocentric outlook" proposes both that "humans are members of the Earth's Community of Life" and that "the human species, along with all other species, are integral elements in a system of interdependence" (1986:99-100). These are the same relational and participatory views which were described in the first chapter of this section, and are perfectly consistent with an eco-holist position.

Taylor's understanding of them is sharply divergent from that of followers of the land ethic, however, and from Native American philosophies. For we also find him asserting that there is "a fundamental duality between our biological nature and our moral autonomy" (1986:48). From this follows the familiar duality between matters of fact and value. According to Taylor, the two must be kept "strictly separate" from one another. Value judgments are not to be derived from matters of fact, even though our biological nature is an enabling condition of our being able to form ethical judgments, and to become moral agents. The one area is, therefore, fundamentally related to the other. Taylor raises these points as explicit criticisms of the land ethic itself:

> The claim is frequently made that ecology shows us how to live in relation to our natural environment. Thus it is held that ecological stability and integrity themselves constitute norms for environmental ethics. The very structure and functioning of the Earth's ecosystems, it is said, make known to us the proper relationships that should hold between ourselves and the natural world. The ecological balance holding among organisms and between organisms and environment in a healthy ecosystem should be our guide in shaping a human culture that fits harmoniously into the order of nature (1986:50-1).

This is certainly a correct, and indeed quite an eloquent description of the eco-holist position, and of its concept of adapting our own patterns of relationship to those which can be observed in the natural world, which he rightly attributes to Leopold and his followers. As Schumacher relates, "the task of the wise man is to understand the great rhythms of the Universe and to gear in with them" (1973:223). Thus *adaptation to natural conditions* is the fundamental norm underlying the land ethic, and this applies not only to our practices, but to the theories which guide them as well. Such a philosophy would also appear to be perfectly consistent with the Native American philosophies discussed above[81].

According to Taylor, however, "[t]his line of reasoning is not

---

[81] Recall especially the discussion in Section I:2

sound from a logical point of view. It confuses fact and value, 'is' and 'ought'" (1986:51). If humans are considered as biological organisms, it is a question of fact, as is a consideration of the ecological context, and of the patterns of interrelationship which form the enabling conditions of those organisms. If we are considered as moral agents then it is questions of value which concern us, such as "How ought I to live?" Since such value judgments cannot, supposedly, be derived from statements of fact due to the "logical gap" which separates them, questions of how we ought to live must be arrived at independently of any facts about how the world actually is.

Consequently, the very idea of adaptation is challenged. For adaptation is just such a modeling of human patterns of relationship (how we ought to live) after those which can be observed in nature (the facts about the world). As Rolston states, "[w]e are searching for an ethic that appropriately 'follows nature.' We want to *optimize* human fitness on Earth (1988:xi, emphasis added). In other words, natural archetypes are essential to the ecological concept of adaptation, and as Narindar Singh adds, "What we need...is a theory that is not only logically consistent...but also *adequate* to reality" (1976:113-4, emphasis added).

According to Taylor, however, the adaptiveness of his philosophy in practice is not the issue. Rather, the test of its validity is the socially derived standards of "correct" reasoning which have traditionally been applied to moral systems in Western philosophy. The *validity* of the ethical system which he proposes, therefore, is to be established *without reference* to the facts of the world, ecological or otherwise. He proposes, instead, five formal conditions which valid moral rules must meet.

The first is that they "must be general in form," and should "not contain any reference to particular persons or actions but only mention *kinds* of things in terms of their properties." Thus, "Thou shalt not kill" would be general in form, while "Frank shalt not kill Anne" would not meet this requirement for validity. The second criterion is that such moral rules "must be considered to be universally applicable to all moral agents," that is, to all human beings. The third requirement is that they must be standards which are "applied disinterestedly." They must be followed out of principle, even if it is not in one's own interest to do so. For example, even killing in self-defense would be prohibited if the moral injunction that "Thou shall not kill" were followed in this manner, since the moral rule is considered to be absolute prior to, and irrespective of, any particular situation or context in which it might be applied. The fourth requirement is that "moral rules must be advocated as normative principles for all to adopt...a moral rule or standard is valid only if those who adopt it as their own normative guide have an attitude of approval toward its being adopted by all others." Thus, "Thou shalt have no other Gods before me" would seem to be an appropriate example here, and European attempts to impose their religious beliefs upon First Nations and other peoples ever since the beginning of the colonial age would seem to be an illustration of European acceptance of this requirement for a valid moral system in practice. The final requirement is that moral rules must also "be taken as overriding all nonmoral norms," so that they may take precedence over mere legal or individual norms (1986:27-31).

Thus, Taylor proposes that the validity of his moral system must be judged against the traditional *a priori* standards of "correct" reasoning of Western culture; against social standards of acceptable logic, or social archetypes. These standards are also considered to be quite certain in advance of any knowledge of the nature of the world in which they are to be applied. In this way, his efforts to validate his ethical system can remain "pure," as it were, untainted by any mention of ecological facts. Rather than adapting his ethic to the natural world, Taylor prefers to adapt it to the trad-itional assumptions, or foundations, of Western ethics–to social rather than to natural archetypes. Indeed, the model of a valid moral system from which he derives his five conditions is that of "respect for persons," that is, the traditional human-centered ethic discussed above. This system was not only human-centered, how-ever, but is also consistent with human superiority to nature, and an exploitive attitude towards the natural world.

Yet while the biocentric view which Taylor proposes is to deal with the natural world as a whole, rather than only with the social sphere, and though he claims to reject all of the above implications of that view, Taylor continues to argue that he can model the form of his own system after that of "respect for persons." They must meet the *same* formal requirements in order to be judged valid. Thus he is *already certain* of the form his ethic must take irrespective of its content or application, and of the nature of the world within which it is to be applied. This implies both that its form can be "sharply separated" from its content, and that he can be certain of it without reference to its adaptive consequences when enacted in practice.

Eco-holism, as always, considers form and content to be fundamentally related to one another, rather than fundamentally separate. It proposes that changes in the content of our ethics imply analogous changes in the form which that ethics will take as well. In particular, it proposes that it is our maladaptive patterns of relationship with the natural world which have given rise to the need for an *ecological* ethic in the first place. Thus, it is upon such patterns of relationship which we must focus. And since our social patterns of relationship are at present inconsistent with natural ones (i. e. are maladaptive), it is to natural archetypes that we must look in order to correct this imbalance, even if this is to derive an "ought" from an "is[82]."

This position also provides good reason to focus upon relational symbolism in our theorizing. For if all things are fundamentally related, as Taylor himself claims, are not facts also related to values? And is not the form a theory takes related to its content and application? To claim otherwise is to once again separate the mind and its "theories" from the body and its "practices." But as Gilbert Ryle so aptly pointed out, "[i]ntelligent practice is not a step-child of theory. On the contrary theorizing is one practice amongst others and is itself intelligently or stupidly conducted" (1949:27). Consequently, when seen as *a part of* its larger practices the traditional standards and patterns of thought of Western culture–its theories–are seen by eco-holists as a consistent part of a larger maladaptive system of practices.

So while Taylor would have us believe that he has adopted relational and contextual views, he seems to have spent too little time considering their implications. He retains all of the traditional dualities of modern Western philosophy, and considers all such pairs-of-opposites to be separate and opposed to one another, rather than complementary and *related*. This fundamental schism runs through his entire argument, and is only made more evident

---

[82] It is also for this reason, I would venture, that Bateson (1988) entitled his last scholarly work Angel's Fear, after the Christian saying, "where angels fear to tread." What I believe he meant to imply by the title was that he would be venturing into territory where scientists have been afraid to tread, since this was the work in which he made his most explicit arguments proposing that ecology provided a scientific basis for ethics.

by that fact that, although he claims to have adopted a relational and contextual world view, his ethic remains completely atomistic. Only by keeping them "sharply separated" could the blatant inconsistencies between his ontology and ethics be so completely and thoroughly overlooked.

Where Regan and Singer propose that the essential attribute which animals share with humans–and which thus makes them worthy of respect–is consciousness, for Taylor it is life. Any living organism which has a good of its own, which is a "teleological center of a life," and which can be benefited or harmed by an action is now considered worthy of our respect. As usual, this position implies a value hierarchy between those lucky beings within the sphere of moral concern and everything which lies without. In this case the division is between individual organisms and the inanimate enabling conditions which make their lives possible. Thus, even though he proposes that "[t]he biocentric outlook precludes a hierarchical view of nature" (as does eco-holism), he also proposes that "the physical conditions of a natural environment must, from a moral point of view, be *sharply separated* from the animals and plants that depend on those conditions for their survival" (1986:45 and 18). As a consequence, it would be more accurate to say that his version of the biocentric outlook precludes hierarchy among those individuals which are seen as having a good, and not hierarchy as such. It is only individual organisms which are considered worthy of moral regard, and not the land itself, nor the natural conditions to which these organisms owe their lives.

Thus, that the parts are fundamentally separate from the whole, and that the life of an individual organism (and its good) are fundamentally separable from vital aspects of its enabling conditions in an ecological context, are both assumed. An organism's life is considered to begin and end at its skin, rather than including the ecological relationships which make that life possible–such as relationships with air, water, soil and sunlight. Taylor's definition of a "natural ecosystem" also misrepresents the standard definition, in that it makes no reference to such inanimate parts of ecosystems. As he states, "[t]he idea of a natural ecosystem as it is to be understood in this book means any collection of ecologically interrelated living things" (1986:3), period. Taylor thus misrepresents the ecological position by conveniently excluding the relationships which such organisms have with their "inanimate" environment from his definition. His definition of a natural ecosystem has been tailored in such a way that it excludes significant aspects of the enabling conditions of all living organisms, which a relational view can only see as *constitutive* of *life itself.* While Taylor's position is again a step in the right direction, therefore, it is neither ecological nor holistic because, quite simply, it is not *relational*, however much Taylor might like to pretend the contrary.

4. THE LAND ETHIC

To the traditional Amerindian, life finds its meaning in the implicit and admiring recognition of the existence, role and power of all forms of life that compose the circle. Amerindians, by nature, strive to respect the sacred character of the relations that exist among all forms of life (Sioui 1995:9).

TAYLOR DOES SHARE one important point of similarity with Leopold and his followers; this being his belief in the intrinsic value of life, which is central to any consideration of ecological ethics. Since their conceptions of what life is differ, however, so also does that which they consider to be intrinsically valuable. So perhaps Taylor's "biocentrism" should be distinguished from the "ecocentrism" of the land ethic, since for Taylor it is only individual organisms which have such intrinsic value. This is because he considers their life to end at their skins.

The land ethic, however, "locates ultimate value in the 'biotic community' and assigns differential moral value to the constitutive individuals relatively to that standard" (Callicott 1989:37). For if life is considered to be an interrelationship among living organisms, and between living organisms and their "inorganic" environment, then it is this interrelationship to which such intrinsic value must be attached, and it is Mother Earth herself–or "the land"–which must be respected. Even if one begins by attributing value to individual organisms or to one's own life, eco-holists argue that if one adopts and adequately comprehends the implications of a relational point of view, this positive valuation must also be extended to encompass the land itself if one's position is to remain consistent.

As Naess describes the implications of a relational view; "[w]e seek what is best for ourselves, but through the extension of the self, our 'own' best is also that of others" (1989:175). So in this view it is not only moral considerability and respect which are expanded into nature, but the concept of the self as well. As a consequence, "[t]he deep ecology movement...asks for the development of a deep identification of individuals with all life forms, and therefore equates increasing compatibility with increased maturity of the individuals" (Naess 1989:85), not unlike the Native American literature. For from a relational point of view our very "self" comes to be seen as a pattern of relationship with the natural environment. And since nature is seen as providing the enabling conditions of one's own existence—that is, since you are a part of it, and it is a part of you—any value attached to the self must also be expanded to encompass the conditions which make its existence possible as well. As Frank Fools Crow expresses this sentiment following Lakota philosophy:

> Our children are us in the tomorrow of life. In them we remain here, and so it will be with their children's children—if the world survives...The linkage of ages has taught us to think in terms of the linkages of many things. Mind, body and spirit are linked together. You cannot consider one without the others...People, other creatures, and the rest of creation are linked together. Thinking in dimensions like these keep us from being narrow and self-centered. Instead, it stretches and expands the mind (1991:62).

In attributing "intrinsic value" to the biosphere, eco-holists are simply returning to a view which Native American peoples have expressed all along—that Mother Earth is sacred, and worthy of our utmost respect. By expressing this idea in the Native American manner, however, we are also able to shed some light on a persistent controversy in the literature of "environmental" ethics. This concerns the question of whether intrinsic value is "subjective" or "objective." Whether it is based merely upon human evaluation (and hence is "in the mind"), or whether it refers to some property which exists in the world itself, independently of this evaluation. For from an eco-holist point of view, even if it is a mental phenomena, it is still in the world. And if one simply states that Mother Earth is sacred then one is less tempted to assume that the term "sacred" must refer to some mysterious metaphysical property which is inherent in the world. Instead, it becomes evident that the question is essentially a matter of the *attitude* which humans take in their relationship *with* the world, and of the implications of various attitudes in practice. For as Fools Crow once replied, when asked whether rock or earth have life or feelings:

> If everything that has been created is essential to life and balance and harmony, then they do. It depends upon how you think and how you define life. If you believe something has life, it has life. *Wakan Tanka* has taught us to think about creation this way, and when we do, the life all things have within them becomes apparent to us, and we treat them accordingly. We do not abuse or misuse them. It is one thing to step on something you think has no life or feelings, and another to step on it when you think it does (1991:66).

Indeed. And if the Earth as a whole is considered to be sacred, so too are all of the beings contained within it. As McGaa relates, also from within the Lakota tradition, "[i]f all things on earth are from Mother Earth, related through her, and sustained from her, there is no basis for prejudice" (1990:29). So once again, as many other Native Americans seem to have realized long ago, a land ethic is necessarily egalitarian, and the intrinsic value attributed to the land is shared equally by all of its members. All things are considered worthy of moral regard and, consequently, all must be respected.

This does *not* imply that we must all become vegetarians, or starve to death so as to "respect" nature, however, for this ethic is not based upon absolutes as in traditional Western philosophies. After all, Leopold himself was a hunter, and often preferred to obtain his meat in the natural way, or to grow his own food. Native American cultures survived principally by hunting, fishing, gathering and horticulture. Thus, they experienced where food comes from first hand, rather than buying Brand Names in plastic wrapped containers at The Store. This provided a more direct, participatory understanding of the fact that killing other beings is a necessary part of life than most contemporary Westerners enjoy, the majority of which are now urban dwellers.

Their respect for the animals and plants was expressed in other ways: such as making use of the entire animal if you kill it, not taking more of anything than one needs and, quite commonly, by ritually apologizing to the animal or plant whose life was taken, and assuring them that it was only done from necessity, as was discussed above[83]. As Fools Crow reiterates:

> Simple living is less wasteful and more in harmony with all other life. Therefore I use earth's gifts sparingly and with gratitude, attempting wherever possible to replace what I consume, and I try to preserve the natural beauty of the earth (1979:202).

To my mind, this is essentially the same sentiment which underlies the popular ecological motto "reduce, reuse, recycle," with an emphasis upon *reduce*. For one of the primary ways in which traditional Native American philosophies and lifestyles have historically expressed their respect for the land has been by not using any more of anything than they needed; that is, by teaching their children to be *non-materialistic*. As Eastman recalls the teachings of his elders, for example, which he received while he was still living the free life of his ancestors as a child, "I was trained to be a warrior and a hunter, and not to care for money or possessions, but to be in the broadest sense a public servant" (1977:1).

From an eco-holist perspective, such attitudes contain an important lesson for the present day, both ethically and politically. As

---

[83] For a more detailed discussion of the rituals and practices expressing respect for animals among the Cree of northern Quebec, see Berkes (1999:83-7).

Naess points out concerning traditional Native American philosophies, "this is a *realistic* egalitarian attitude, an acknowledgment of the cycles of life and their interconnection in nature" (1989:176, original emphasis). Once again, life and death are seen as forming a cycle, just like other causal cycles in an ecosystem. And a "realistic" naturalistic egalitarianism (or natural democracy) recognizes that death cannot be eliminated, but is an integral part of the larger processes of life.

Egalitarianism, unfortunately, is a point on which even Rolston fails in his holistic vision. Instead, he imports the relational symbolism of hierarchy and human superiority into his larger account of ecological holism. Considering the fact that Rolston proposes to "follow nature," it should not be surprising that this adoption of hierarchical symbolism is based upon a misconceived ontology. For hierarchical symbolism is everywhere in Rolston's model of natural patterns of interrelationship, principally in his many references to the "trophic pyramid."

What he means by this term is that those animals at the "top" of the food chain eat those which are "lower down" the pyramid, with each lower trophic (or nutritional) level being more numerous than those which prey upon them. And of course there is a *sense* in which this is a true picture of ecological relations, if only an incomplete one.

For example, when biocides such as DDT are sprayed in lakes they are first absorbed by algae and plants, and then passed on to fish and other animals which eat them. Being persistent and fat soluble chemicals, as they are passed "upwards" such toxins tend to accumulate in greater concentrations in the body fat of those beings closer to the "top" of the "trophic pyramid." This includes the humans who eat the fish who eat the plants or, as in the case of

Lake Apopka in Florida, fish-eating birds (Williams 1999).

This process is known as biological amplification, since toxins increase as they are passed along the food chain, and concentrate in the process because the "higher" levels consume many of the "lower[84]." What Rolston's account fails to inform us of, is the fact that this "pyramid" is in reality only one half of a *causal cycle*, and the manner in which the "highest" forms are, in their turn, made use of by the "lowest."

Such a "trophic pyramid" is, however, also a perfect reflection of the classic Aristotelian hierarchy ranking humans above animals, which are above plants, which are above the many inorganic parts of nature which sustain the lives of all living things–such as sunlight, clouds, rain, rivers, soil and yes, even the rocks which erode to provide the material for soil formation, and without which terrestrial animals would have no place to live. Furthermore, the Aristotelian hierarchy was modeled after the pattern of use which the "trophic pyramid" describes. In some strange way, the simple fact that one organism makes use of another, as when it eats it, for example, was then seen as a demonstration of its superiority.

Yet if all that is required in order to illustrate one being's superiority to another is the fact that the one makes use of the other, then humans are inferior to the lice which infest their hair, the worms which sometimes inhabit their guts, the mosquitoes and ticks which nourish themselves upon their blood, and to the trees, microbes and insects which make use of their bodies after they die. Thus, the *complete* pattern of use does not describe a pyramid, with

---

[84] Biological amplification also provides a good ecological reason for not eating too much meat (as many North Americans do), since cattle, pigs and poultry are higher up the food chain than grains and vegatables, and therefore are contaminated with higher concentrations of persistent organic pollutants (POPs), a significant percentage of which are carcinogenic (McGinn 2000).

a *linear* pattern of exploitation of the "lower" by the "higher," but rather form themselves into a circle, with a *cyclical* pattern of use. Humans feed upon cattle, feed upon grass, feed upon soil, which is derived, in part, from dead humans. By following hierarchical logic in this situation we can soon end up superior to ourselves[85]!

The more adequate model appears to be that of a food *web*. As Bookchin observes, "ecosystems cannot be meaningfully described in hierarchical terms...The web can be entered at any point and leads back to its point of departure without any apparent exit" (1991:26). After all, the "top" of this pyramid also dies and returns its substance to the Earth, from which other lives then grow. So the fact that humans are not eaten by wolves or other predators very often does not imply that life does not make use of us–or that we are not a part of the circle of life–quite the contrary. We are simply like those predators themselves; the "highest" is often made use of by the "lowest" in the end.

Rolston does not model his hierarchy only upon this pattern of use, but adopts a position very similar to that of Kant. As he states, "[e]ach natural kind has place, integrity, even perfections, but none of the others reaches the eminence of personality" (1988:68). He proposes that there is a hierarchy of intrinsic values based upon the capacities of various organisms–from the inorganic, to the organic, to the conscious, to the self-aware, as with Aristotle. Thus, Rolston feels that there are "higher" goods possessed by humans which animals and other beings do not share. As an example he points out that "[a]n elk and a human both have interests in eating, but an elk has no interest in learning to read; and a human does" (1988:74).

From this it should be evident that what is of value is relative to the good of the organism which it serves, from which it follows

---

[85] Which may be the point, politically.

that reading is valuable to humans but not to elk, to whom reading has no value whatever. For even our much vaunted reason, which Rolston places at the top of his hierarchy of intrinsic values, is most often of value only to ourselves. Indeed, the way in which the Western world has been employing its reason of late has proven to be a *disvalue* to most other forms of life, *and even to the possibility of the perpetuation of our own species*. It seems there may be some empirical evidence to suggest that reason has been somewhat overrated, at least in its traditional Western forms, and that its superiority cannot be assumed in advance of an examination of Western practices. Yet Rolston proposes that "[a] human who can eat and take an education has more interest in eating than does an elk, since all the 'upstairs' values depend on the 'downstairs' ones" (1988:74).

What I fail to understand is how it follows from the mere fact that humans have different *capacities* or *talents* from those of elk, or that they have all of the capacities which elk have and then some, that they are superior. That humans *do* have different capacities is certainly true–particularly the ability to reason by means of a highly complex symbolic language–and from this it *does* follow that we have different abilities, perhaps even more than those which elk have. For reason does open up options which are unavailable to other animals–yet so do wings, antlers or gills. And it does not follow from the mere fact of difference that one of the differentia is therefore inherently superior to the other. Red is different from green, yet questions of superiority and inferiority are irrelevant, or at least remain to be demonstrated with reference to a particular situation (green being better camouflage in the woods, for example).

This is something which Rolston, and the tradition which he

follows, never seem to have done with human reason. And since reason seems to be an evolutionary experiment whose ultimate value as an adaptive capacity is currently being tested by the ecological crisis, perhaps judgments as to "upstairs" and "downstairs" values had best wait. Abilities which are valuable to humans are, after all, *human* capacities, and are relevant only when considering the good of humans. Other animals have other goods to which other capacities are the appropriate means of fulfillment, and there are many things which contribute to the good of all life. Yet the value of any *capacity* remains relative to the good which it serves, and these goods are relative to the type of organism being considered. Indeed, because reason allows us to value the Earth as a whole in a way in which no other organism can, humans also have the *potential* to contribute toward the good of all life in a way in which few other single species could. As an anonymous Onondaga saying puts it:

> Being born as humans to this world is a very sacred trust, we have a sacred responsibility because of the special gift we have, which is beyond the fine gifts of the plant life, the fish, the woodlands, the birds, and all other living things on earth, we are able to take care of them[86].

It does not follow that we do so. And since the only way in which human reason can prove its worth in the present age is to put itself in the service of the good of Mother Earth, any ranking of these capacities in hierarchical order is simple anthropocentrism at present. By turning to the hierarchical relational symbolism of the human-centered position which he rightly opposes, therefore, Rolston obscures the difference between their position and his own. His insistence upon human superiority based upon the superiority of reason is precisely the argument used by those for

---

[86] Quote from a page from an old calendar of mine, currently on my bathroom wall, above the towels. I have never been able to track down the original source.

whom nature was merely "waste" unless it was made use of for some economic purpose, and in which nature need be afforded no ethical regard whatever, existing merely for the use of humanity. Since these are the implications which most of his readers will bring to this position, his fear of denying human superiority–of denying that the Earth is the center of the universe, and may yet revolve about the Sun–does not help to clarify the situation.

Consistent with the relational symbolism deployed by Bateson, Lame Deer and many other Native American thinkers, this traditional hierarchy might best be transformed into a "holarchy." That is, into a series of four "circles within circles," which describes these various qualitatively distinct realms of being in terms of the manner in which they are interrelated to form a larger whole. As Schumacher puts it, "[o]ur task is to look at the world and see it whole" (1977:15), not to chop it up according to its perceived value from the human point of view. In this case, the largest circle–the most complete context–would be that of *"non-living"* beings. Inside of this would be located the subcategory of *living* beings, then *sentient* beings, and finally, *self-aware* beings such as humans. In this way we see that differences need not imply superiority and inferiority at all–as they generally *do* in social hierarchies–instead, differences are merely differences.

The outer ring is the physical universe; the so-called "inanimate" world. Yet in terms of the *evolution* of these categories of being, it is the enabling condition for all the smaller, less inclusive categories which evolved within it[87]. Any value attached to them must be attached equally to it, since they could not exist, nor carry value, unless this larger context did. This is similarly true of each of the progressively smaller, less inclusive spheres–from life, to

---

[87] And the same applies to the evolution of the ecological, social and meaningful patterns of relationship in the holarchy proposed above.

sentience, to self-awareness. Each larger sphere is a precondition of those within it. Each smaller sphere could not exist except that the qualities described by the larger ones existed first. Not only are our lives continually made possible by the air, water, soil and sunlight of the largest sphere, and by the plants and animals of the next smallest spheres, so also each of these smaller spheres could only have evolved within the context provided by the next largest. As Eastman observed, "I have sometimes wondered why the scientific doctrine of man's descent has not...apparently increased the white man's respect for these our humbler kin" (1980:143). For just as life evolved from inorganic matter, so consciousness evolved from life, and self-awareness from consciousness. Each is simply a more specific type of all of the former, from which its actuality was *and is* derived.

An ecological ethic, then, is not based upon the model of social hierarchies, but upon an understanding of the larger world to which that ethic must be adapted. And humans, with their self-awareness, are no more valuable in any ultimate or final sense than the "dead" rain without which their lives (and whatever values are wrapped up in them) are impossible, let alone more valuable than the grass, the trees, the whales or the deer. Each has a right to live, and each must be respected in a manner appropriate to its kind, *especially by allowing its kind to go on living in its accustomed manner*[88].

This does *not* preclude the fact that one organism may still make use of another. For it is perfectly natural for all living things to make use of their environment, not because they are superior to it (which I doubt could even enter the consciousness of any organism but a human one), but because such use is a necessary *part of their life*. Quite simply, the wolf does not consider the deer to

---

[88] This statement has not only ethical, but also political implications, as shall be discussed below.

be *inferior*, but it still eats it.

So while hierarchy is not necessary in order to justify the use of the natural environment–but only necessity itself–hierarchy *is* useful in justifying the domination of nature and a total unconcern for the implications of human actions upon it. An egalitarian view, on the other hand, implies a more respectful and restrained use of the gifts which Mother Earth provides.

The above discussion also suggests another persistent difficulty in environmental ethics which a relational view leads us to reassess. This is the opposition between intrinsic and instrumental values. For unlike Locke's distinction between things which are useful to the perpetuation of life and things which are profitable (which was discussed in Section II:4), contemporary writers tend to follow the Kantian ethic, in which intrinsic value is attributed to specific metaphysical properties or capacities inherent in individual beings.

For example, the human rights movement attributes intrinsic value to all persons on the basis of the fact that they are self-aware or can "say the word 'I'." The animal rights movement attributes intrinsic value is attributed to all conscious animals. Taylor attributes it to all living organisms. In each case, however, it is to the metaphysical properties inherent in individuals that such intrinsic value is attached, while their relationship to the larger world is not encompassed by this positive valuation.

In this way the *intrinsic* value of the individual, or whatever metaphysical property it contains as a subjectivity, may remain "sharply separated" from the *instrumental* value attributed to the natural conditions of which it makes use, and which make its life possible. For as Kant put it in his categorical imperative, such individuals must universally be treated as *ends-in themselves* and *never merely as a means*.

While this is certainly an admirable and appropriate sentiment when applied to inter-human ethics (with the question of whether it is consistent with wage labour being set to one side for the moment), it is totally inadequate when applied to the land. For it is only when an organism's life is considered to end at its skin that we can clearly distinguish between instrumental and intrinsic value in the senses described above. In that way, following the usual dualistic metaphysic, the intrinsic value of its life (subjective, or skin-in) can be separated from the instrumental value of its relationships with its context (objective, or skin-out).

Yet when life itself is defined as such a pattern of relationship, as it is by eco-holists, this distinction is no longer tenable. For if the life of an organism has intrinsic value, and if its life is a pattern of relationship to *and use of* its environment, then these instrumental uses, or enabling conditions, also have intrinsic value. After all, life itself is just such a pattern of use between an organism and its ecological context–a pattern of instrumental use without which the organism could not exist, nor, therefore, its intrinsic value. Consequently, when the organism–or the self–become contextualized, instrumental values are no longer "secondary," pursued only as a means of perpetuating a previously existing intrinsic value, or a life. Instead such instrumental values are *a part of* that life, or an enabling condition of it. Intrinsic value can no longer be confined within the skins of individuals in a relational view. It necessarily consists in this relationship–this pattern of use–between an individual and its ecological context. Thus, as Rolston observes:

> There is nothing wrong with humans exploiting their environment, resourcefully using it. Nature requires this of every species...But humans have options about the extent to which they do so; they also have, or ought to have, a conscience about it (1988:158).

Or as Naess puts it, "[h]umans have modified the Earth and will

continue to do so. At issue is the nature and extent of such inter-
ference" (1989:30). At issue is also the *means* by which we decide
which modifications are desirable. This has generally been decided
according to *economic* criteria in recent times. So of more relevance
to a land ethic than the intrinsic/instrumental distinction of the
individualistic position is a distinction closer to that of Locke. This is
a distinction between two types of *instrumental* value–that is,
between uses which are *intrinsic* or *extrinsic* to the *maintenance and
perpetuation of life in general*. Or again: between a things usefulness
to the perpetuation of life in general–its *intrinsic* use value–and its
usefulness to the production of a profit, a type of use value which is
entirely *extrinsic* to the direct maintenance and perpetuation of life.
And it is precisely *this* contrast which Leopold chose to focus upon
in his original formulation of the A-B cleavage, which we have
come to know as the contrast between shallow and deep ecology
in the above discussion. As Leopold put it:

> It is inconceivable to me that an ethical relation to land can exist without
> love, respect, and admiration for the land, and a high regard for its value,
> and I of course mean something far broader than mere economic value...The
> 'key log' which must be moved to release the evolutionary process for an
> ethic is simply this: quit thinking about decent land-use as solely an
> economic problem (1949:223-4).

Or as Rolston phrases it, "[t]he forces that drive capitalism need
moral watching...We want our ethical attitudes to be consistent
with ecology, not distorted by economics" (1988:158). Nor were
Native American philosophies "distorted by economics," since up
until their contact with Europeans most *had no money*, and had not
yet fallen under what Standing Bear once described as "the
contaminating influence of the white man's dollar" (1978:184). Thus
they could not help *but* value things for their own sake, or as a
means of continuing to lives themselves, rather than as a means of

making a monetary profit. This is one of the key reasons that eco-holists turn to them as models for the development of more ecological sensibilities, as shall be discussed further below.

So the contrast is quite simple: between the value of life, and that which is needed for its perpetuation, and the value of money. And the land ethic points out that a great many other values override economic ones. Of central concern is the value of the land itself, which is considered to be more fundamental than the money values generated by economic activities. After all, the latter may be quite profitable even while they are dangerous to all life–as with nuclear power and many toxic chemicals. In fact, the activity of economics involves adapting the ecological context to the short-term purposes of humanity, rather than humans to their ecological context. The end which economic activity seeks is nothing more than the satisfaction of human preferences by adapting natural resources to the ends which humanity chooses. And since adaptation in the ecological sense is not even its aim, it is unlikely to achieve it left to its own devices.

What is required in order to address our current ecological problems is a more basic standard of value based upon a more adequate description of the values actually resident in nature. One which can encompass values more fundamental than an activity's profitability or an object's monetary worth. This is the value of the perpetuation of life on Earth–of life itself. And this standard of value arises from values, or *valuings*, in nature itself. Rather than arising exclusively from social conventions, it arises from the *relationship* between ourselves and more-than-human nature.

As Rolston relates, "an organism is a spontaneous evaluative system" (1988:102). In defending its life, in seeking certain ends, and in making use of its environment, an organism demonstrates that it *pursues certain values*. The roots of a tree seek water, which is of value to it. The moose seeks forage, which it values. A wolf seeks its prey, which it values. All of these things are valued as intrinsic use values, or as a means of perpetuating the good of the organism itself–its life. And as Okute adds, "[a]ll living creatures and all plants are a benefit to something" (as cited in McLuhan 1971:19).

From an evolutionary perspective, such *valuing* on the part of living things was also an enabling condition for the development of *evaluation* on the part of humanity. *Evaluation*, one might say, is the linguistic and conceptual form of valuing which is peculiar to human beings, while *valuing* is a talent which they share with all other living organisms[89].

Thus, evaluation becomes a particular (human) instance of this more widespread ability common to all organisms, and from which human evaluation evolved–the ability of all organisms to distinguish between, and pursue, certain values in the context which surrounds them. Among non-human beings, such values are inevitably *intrinsic* use values–they pursue only those things without which their lives could not be maintained. Consequently, this sense of intrinsic value is best understood as a verb–as a *valuing*. And if this valuing is observable in nature, rather than being something which is peculiar to humans, then our own patterns of evaluation may also be patterned after those patterns of valuing which are empirically observable aspects of the natural world. We may, in other words, pattern our evaluations after natural archetypes, or after the values which all life pursues. So to return to the teachings of Black Elk, "nothing can live well except in a manner which is suited to the way the Power of the World lives and moves to do its work."

---

[89] I would like to thank Dr. Peter Miller for suggesting the distinction between valuing and evaluation as I use them here.

In being based upon natural archetypes rather than social ones, this ethic is more likely to result in adaptive policies when enacted in practice. For the value which we attribute to money–to little symbolic pieces of metal and paper, or to the bit of ink or data which records our current balance at the local bank–is based purely upon human social conventions. Anyone who doubts this should try to spend it while they are alone in the bush. The value which we attribute to life, on the other hand, and which we ought to attribute equally to the Earth which sustains that life, is a valuing which humans share with all other living organisms.

And it is this type of valuing which is most required of us now. For as Asa Bazhonooday, an elder of the Navaho people, has encapsulated the point in question following the traditions of her own people:

> The Earth is our mother. The white man is ruining our mother. I don't know the white man's ways, but to us the Mesa, the air, the water, are Holy Elements...How much would you ask for if your Mother had been harmed? There is no way that we can be repaid for the damages to our Mother. No amount of money can repay, money cannot give birth to anything (as cited in Nabokov 1991:399).

IV

THE CLOSING CIRCLE II: ECOLOGY & INDIGENOUS PEOPLES IN THE CONTEXT OF GLOBALIZATION

1. GLOBALIZATION AND NEO-COLONIALISM : THE VIEW FROM POLITICAL ECONOMY

> the economic incorporation of small-scale cultures into the world-market economy is a critically important phenomenon. A tribal culture may surrender its political autonomy but can still continue to be an essentially small-scale culture if it is allowed to retain its self-sufficient, subsistence economy and if it remains unexploited by outsiders. The degree of a small-scale society's participation in the cash economy must be determined by the people themselves on their own terms. Only in this way can economic self-sufficiency and cultural autonomy be safeguarded and the 'price of progress' minimized (Bodley 1999:111).

THE PRESENT SECTION shall turn to a discussion of the implications of an eco-holist philosophy for the theory of practice, especially in the contemporary political and economic context. The emphasis shall once again shift from a focus primarily upon the ideological sphere to include a consideration of practices in the social and ecological contexts or, more specifically, to what is often referred to as political ecology. This will include a discussion of eco-holism's critique of the trend towards economic globalization, and its con-sequences, which has characterized the expansion of the capitalist (or modern) world system since the early colonial age. The manner in which these critiques are consistent with the previous discussion of Native American philosophy will be returned to throughout.

From an eco-holist perspective (or from a scientific one for that matter) it is always by looking to nature that such ideological debates ought to be settled. This is because basing one's ethos and world view upon an adequate understanding of the nature of our relationship with reality is a necessary prerequisite of adaptation–its fundamental goal in practice. Eco-holism judges the adequacy of social systems, technologies and even ideologies relative to this standard. As Rappaport observed:

> The criterion of adequacy for a cognized model is not its accuracy, but its adaptive effectiveness...in this way it becomes possible to assess the adaptiveness not only of overt human behavior, but even of the ideology which informs that behavior (1979:98).

Of central concern in the present section shall be the maladaptive expansion of economic and political patterns typical of the West, the spread of Western technology and the ecological effects with which it is associated, a critique of the philosophies consistent with these patterns of development, and the common themes crosscutting indigenous and ecological responses to the same in the contemporary world.

The first theme, an emphasis upon the importance of appropriate cultural scale, originated within alternative economics as a critique of the dominant trend towards centralization. As John Bodley (2001:235-6) suggests, the economist Leopold Kohr (1978) was among the first to emphasize the relevance of appropriate cultural scale in his book *The Breakdown of Nations*. While Kohr's work largely remained in obscurity, the central thesis concerning appropriately scaled patterns of human organization was popularized in the works of E. F. Schumacher (1973, 1977,1979, 1997), who is now widely considered to be the father of Green economics[90].

---

[90] Esteva and Prakash (1997:279) also note the affinity of appropriate scale theory with the earlier thought of Mahatma Gandhi, one of the key leaders of India's independence movement, and Schumacher's ideas remain popular with many Indian academics.

While the importance of cultural scale is central to Bodley's writings, many of his ideas are inspired by earlier anthropological literature on political economy, which began to enter the discipline in the 1970s. Consequently, many of his critiques of the capitalist system are consistent with those of political economy more generally, and of the various understandings of the capitalist world system inspired by it, including dependency theory and world systems theory[91].

The present discussion will examine the contiributions of the political economy literature to these issues before developing upon the cultural scale critique in more detail.

As Roseberry observes, historically:

> Political economy could...be distinguished from neoclassical economics, which represented a shift in concern from the 'wealth of nations' to the price of beans, from value as determined by labour time to price as determined by markets (Roseberry 1988:162).

Thus, political economy's more holistic emphasis upon the study of systems of production has largely been reduced by neo-classical economics to a study of the pricing of commodities–the study of markets and market values. Further, the fragmentation of its area of study into the disciplines of economics, political science (the study of political systems) and sociology (the study of social organization) is a further example of the institutionalization of an atomistic premise within Western academic institutions. Rather than viewing politics and economics as separate and opposed, political economy saw them as part of an integrated system of production and distribution, a much more holistic position. Given anthropology's own holistic approach, it was sympathetic as a discipline to a political economy approach, though as Roseberry

---

[91] For an excellent overview of the political economy literature and associated approaches see Roseberry (1988).

suggests, anthropologists did not appropriate the early political economy literature as a whole, rather, "[t]hey appropriated Marx" (1988:162).

One of the most central concepts of Marxist thought adopted by the recent political economy literature is that of *modes of production*, which describes the manner in which labour has been deployed differently in different types of societies, and which social class (if such exist) controls access to the means of production. The emphasis is upon neither politics nor economics, but rather upon the role which *both* play in the organization of production systems, the allocation of labour and the distribution of that which is produced (wealth) among the members of society. As Lewellen suggests:

> Production is not simply a matter of people using nature to create goods; more importantly, production–which includes labour, technology, ownership and transportation–determines the ways that people relate to each other, that is, the ways societies are organized (1992:162).

Ownership of the means of production is a central concern. Eric Wolf (1982), in particular, has popularized the centrality of the concept of modes of production. Wolf proposed that there have been three different modes which have characterized human societies historically, "a capitalist mode, a tributary mode, and a kin-ordered mode" (Wolf 1982:76)[92].

A mode of production is often defined as including both the *forces* of production and the social *relations* of production. The

---

[92] Wolf's (1982) book, *Europe and the People without History* is widely considered to be a ground breaking work within the political economy literature. See, for example: Lewellen (1992:161-66) and Roseberry (1988:173). These distinctions were foreshadowed, however, in the work of Marshall Sahlins, who suggested that "The household is to the tribal economy as the manor to the medieval economy or the corporation to modern capitalism: each is the dominant production institution of its time. Each represents, moreover, a determinate mode of production" (1972:76-77).

forces of production include such factors as technology, labour power, skills and knowledge. The social relations of production, on the other hand, include such processes as the organization of work, the institutions by which labour is allocated, class relations and the manner in which access to the means of production and the distribution of its products are organized (Bates and Fratkin 1999:50).

A kin-ordered (or domestic) mode of production was typical of all pre-state societies, such as those of the Native American cultures discussed above, often described as tribal societies. In general, such societies are relatively egalitarian in their distribution of wealth, labour is allocated within households and larger groups based upon kinship and marriage ties, and the land is considered to be a collective inheritance or responsibility. The fact that everyone has equal access to the means of production, coupled with social norms emphasizing reciprocity within the household or wider kin group, create a relatively equitable division of wealth in a kin-ordered mode, or a form of substantive equality. Both the production and distribution systems, as well as the organization of labour, are largely organized around considerations of kinship, and this mode "is based upon an opposition between those who belong to the group and those who do not" (Lewellen 1992:162). A kin-ordered mode is, therefore, a *classless* society, in contrast to the other two modes. As Wolf relates:

> both the tributary and the capitalist modes divide the population under their command into a class of surplus producers and a class of surplus takers. Both require mechanisms of domination to ensure that surpluses are transferred on a predictable basis from one class to another...Both the tributary and the capitalist modes, therefore, are marked by the development and installation of such an apparatus, namely the state (Wolf 1982:99).

What both modes share is a stratified, class system in which an elite control access to the means of production and expropriate surplus from the lower class, though the means vary. A tributary mode was typical of ancient states, such as Egypt or Mesopotamia, as well as the feudal systems of Europe, China and Japan[93]. In this mode, a noble or priestly class control access to crucial resources, such as land or irrigation works, and use coercive political institutions to extract taxes to support themselves, their retainers and the state apparatus. As Wolf describes it, "social labour is...mobilized and committed to the transformation of nature primarily through the exercise of power and domination–through a political process" (1982:79-80). The "surplus" is usurped through taxation, with such taxes being levied in the form of tribute (a share of agricultural production) and/or in terms of corvee labour (mandatory state labour). A tributary mode is thus a highly stratified system in which an hereditary noble, priestly or "landed" class control access to the means of production, most often by claiming hereditary ownership of the land, and using political coercion to extract surplus produce and/or labour. The classes, in this case, are *closed*, in the sense that there is little chance for movement between them under normal circumstances.

The capitalist mode of production, on the other hand, began with the collapse of European feudalism after 1500 A. D., coupled with the initial expansion of European trading and the extension of colonial control. It then flowered in the industrial revolution of the 1800s, and matured in the 20th century into an increasingly global system, which has come to incorporate virtually all other previously autonomous polities and economies (Bodley 1999:6). In this mode. an "open" (or at least theoretically open) class of very wealthy individuals has come to control access to a much more technologically complex system of production, primarily by means

93. Here Wolf conflates two modes of production between which Marx had distinguished, the Asiatic and the feudal (Lewellen 1992:162).

of private or corporate ownership. True access to this class is achievable only through economic competition at the grandest scale.

The "surplus" extracted in this mode is not a share of production, but rather, as Marx suggested, the *surplus value* extracted from labour. As Lewellen makes this point, "[t]hat which workers produce above the value of their wages is a surplus transferred to the owners" (1992:162) of the means of production–the capitalist class. While state guarantees of private property are necessary for the system to function, it is primarily by means of "economic compulsion" (Bernstein 2000:261) that surplus value is extracted. And for economic compulsion to work, the means of production, from land, to markets, to factories must be privately owned. As Wolf relates:

> As long as people can lay their hands on the means of production (tools, resources, land) and use these to supply their own sustenance–under whatever social arrangement–there is no compelling reason for them to sell their capacity to work to someone else. For labour power to be offered for sale, the tie between producers and the means of production has to be severed for good (1982:77).

While Bodley admits that Wolf's three modes of production "correspond broadly" to the three cultural scales which he identifies, the emphasis is somewhat different (2001:55). Where Wolf categorizes societies into modes of production based upon the allocation of social labour within the *production* process, Bodley focuses upon social organization more generally, and the *scale* at which societies are organized. Consequently, Bodley "sort[s] cultures by their dominant organizational forms, proposing a distinction between domestic, political and commercial scale cultures" (2001:13). Where Wolf focuses upon classes and their relationship to the means of production, Bodley suggests an emphasis upon the "dominant cultural processes" at each level of social organization, the values which they pursue (2001:225), and the relationship between "cultural scale and social power" (2001:16). As Bodley puts it himself:

> Societies have developed three ways to organize culturally the distribution of social power and material living standards: (1) domestically by means of the household, (2) politically by means of a ruler, and (3) commercially by means of the market. Each approach seems to require a different scale of society and produces a distinctive distribution of social power and household living standards (1999:7).

In domestic scale cultures, social organization is organized entirely by households and kin groups at a *local* level, and "households, as the minimal unit of society, are largely self-sufficient and highly autonomous" (Bodley 2001:14). Social power is widely dispersed in domestic scale cultures. Bodley describes the dominant process or goal of social organization in domestic scale cultures as *sapienization*; the reproduction, development and maintenance of human beings, as well as human cultures and societies (Bodley 1999:4, 2001:3).

Because they are domestic scale cultures, "[t]he focus of the sapienization process is necessarily to promote the well-being of the household" (Bodley 2001:13), and by the extension of kinship principles, the well-being of the *community* of households–or society–of which one's own household is a part. As Cajete describes the process with reference to Native American societies, "[p]eople realized themselves by being of service to their community and by caring for their place. They sought the completion of themselves as tribal men and women" (2000:95).

Bodley's cultural scale emphasis is meant to suggest that many contemporary human problems–both social and ecological–have been amplified as the scale of human organization has increased. In addition, the goal of "sapienization has been super-seded by two other cultural processes...that have undermined the well-being of many households while bestowing enormous ad-vantages on others" (2001:13).

The first of these processes is the *politicization* process. It was associated with the rise of tributary states–or in Bodley's termin-ology, political scale cultures. In such societies, social power came to whichever class controlled the political apparatus of the state, and access to land. The second process is *commercialization*, which gave rise to the capitalist mode of production–or in Bodley's terminology, commercial scale cultures (2001:4). In such societies, social power increasingly came to be concentrated through the commercial ownership of the means of production.

The former increased the scale of social organization from a local to a regional scale, through the incorporation of formerly autonomous agricultural villages into tribute paying peasantries. Political centralization and state coercion were the means. The latter has increasingly incorporated the entire world into a single, global market economy, through the globalization of commercial activity and the creation of economic compulsion. Where in the former political processes and values came to predominate over those of the domestic scale, in the latter commercial processes and values came to predominate over both domestic and political scale organization and values.

Besides centralized organization and the state, both political and commercial scale cultures also tend to share another basic characteristic–both tend to be *expansive*. Unlike domestic, or small-scale cultures, which tend to be relatively stable, large and global scale cultures tend to be "disequilibrium cultures." As Bodley suggests, "[a]ncient civilizations were inherently expansive systems which were characterized by frequent institutional collapse" (1999:5)[94].

The expansion of political scale cultures historically, was limited by the available communications and transportation technology. Such societies were also limited to areas of the world which were suitable for intensive agriculture. Commercial scale culture is also inherently expansive. As Wolf relates, "the internal dynamic of the capitalist mode may predispose it to external expansion" (1982:79). Just as the expansion of ancient states involved the incorporation of previously autonomous intensive agricultural communities into states, the expansion of commercial scale cultures has involved the incorporation of virtually all societies into a common market. As Bodley suggests, "[c]ommercial forces have surpassed both sapien-ization and politicization as dominant cultural processes in the world" (1999:6).

What this implies is a colonial process through which previously autonomous domestic and political scale cultures were incorporat-ed into the world system, most often involuntarily. This process has been examined from both the political economy (Bernstein 2000; Robbins 1999; Wolf 1982) and the cultural scale perspectives (Bodley 1999, 2001) in the anthropological literature. Wolf, for example, focused on "the spread of the capitalist mode and its impact on world areas where social labour was allocated dif-ferently" (1982:76). Bodley (1999) provides a detailed discussion of the methods–often violent–by which non-Western peoples, and tribal peoples in particular, have been (and continue to be) in-corporated into the world system. This expansive dynamic began in the early colonial age, and was based upon an expansion of long distance trade in the 16th century, followed by the imposition of

---

[94] For a discussion of the ecological factors involved in the collapse of ancient states, and a comparison with the contemporary situation, see Weiskel (1989).

direct political rule and colonization over vast areas of the globe in the centuries which followed[95].

While the colonial age ended, officially at least, when the European powers granted political independence to the vast majority of their former colonies (mostly in the post-World War II era), essentially the same expansive process continues today, though it is now often glossed as "globalization." Bernstein draws a contrast between "colonialism," based on direct political subjugation, and the concept of "imperialism," or domination by extra-political means (2000:250)[96]. Similarly, Teeple suggests a contrast between colonialism and "neo-colonialism...the subordination of former colonial countries mainly by economic means and principles." The latter was "established as the new relation between industrial metropolises and the 'lesser developed' nations after World War II" (2000:57)[97].

Where the former, or "early colonialism," used direct political coercion to achieve expansion, the latter, "late colonialism," uses economic compulsion. Where early colonialism was associated with slavery and other forms of forced labour in some areas (Bernstein 2000)[98], late colonialism is characterized by what Arthur Solomon, an Ojibway elder and teacher, describes as "a wage slave economy...that is different only in degree to the slavery that was long ago" (1994:67).

---

[95] For more detailed discussions of the phases of colonialism and capitalism, and the relationship between them see Bernstein (2000:245-52) and Robbins (1999:1-140). For an excellent discussion of colonial administrations in Africa and Indian see Potter (2000).

[96] Bernstein derives this sense of the use of the term "imperialism" from a pamphlet of the same name by Lenin, originally published in 1916.

[97] Bodley also makes an implicit contrast between "the colonial period" and "economic development...a special category of cultural modification" (1999:93).

[98] Bernstein provides a detailed discussion of the labour regimes in both the early and late colonial systems.

In order for economic compulsion to function properly, at least three of the preconditions for capitalist expansion must be established. Both Bernstein (2000:261) and Teeple (2000:140-1) describe this process as "proletarianization," which involves the creation of a landless working class.

The first predondition for economic compulsion is the *imposition of centralized modern states* in place of previous forms of political organization, which has happened on a global scale. Thus, the globalization of hierarchical patterns of social organization is the first lasting legacy of the early colonial period. The state, from a capitalist perspective, serves the function of: (1) guaranteeing private property, and (2) promoting the other two processes. For as Teeple suggests, "most of the former colonies were, under the guise of political freedom, opened to international trade" (2000:57).

The second process is the *commodification of lands and resources*. As discussed above, colonization has always sought not only to control other peoples, but to control and exploit their lands and resources as well. And as Bodley suggests, "tribal cultures and their environments have been devastated as the resources they controlled have become necessary components of commercial growth" (2001:91).

Communal property rights regimes, such as those of indigenous peoples, have been systematically dismantled on a global scale, because they stand in the way of colonial expansion and "progress." Thus, around the world, "[s]trenuous efforts were made to destroy indigenous communities by forcing them to divide up their common lands and hold them in individual parcels" (Maybury-Lewis 2002:5); a pattern more consistent with the dominant individualistic social philosophy of the West, and with capitalist expansion. As Standing Bear relates, "[i]t was land–it has ever been land–for which the white man oppresses the Indian and to gain possession of which he commits any crime" (1978:244).

The settling of the Americas marked the beginnings of European expansion. Yet "while the Americas furnish the oldest and most dramatic example of the treatment of indigenous peoples in this regard, similar processes have taken place all over the world" (Maybury-Lewis 2002:5). Indeed, common property regimes have widely been viewed as an impediment to "development," or even as inherently unmanageable on a sustainable basis (Hardin 1968). The commodification of tribal or communal lands has left most indigenous peoples as marginalized populations in their own traditional territories. As Maybury-Lewis concludes, "the development strategies pursued by state governments have deprived tribal peoples of their lands and resources and reduced them to poverty or actual starvation" (2002:34) in many cases.

Communal rights to land have never been properly recognized or protected by the international community nor, therefore, have the communal rights of indigenous peoples. Ironically, perhaps, the common ownership of transnational corporations by various shareholders has never been questioned, thus implying a clear double standard when it comes to dealing with Western and non-Western understandings of common property in practice.

As a result, both the ongoing commodification of lands, and the ongoing colonization of indigenous resources continues to be promoted by states through a type of *internal* colonialism[99] even to this day. As Bodley suggests, "the cultural modification policies of independent modernizing nations have sometimes been more sweeping in their destruction of tribal culture than were the earlier colonial authorities" (1999:96).

---

[99] Lewellen defines internal colonialism as "a situation...in which a few small elites centered in one city exploit the rest of the country" (1992:155).

The third process, or precondition for the establishment of economic compulsion, is the *monetarization or commercialization of all non-capitalist markets and production systems*. The near universal-ization of money as a common standard of value, and the inter-nationalization of the market economy is the third lasting legacy of the colonial process. This involves a further commodification of economic products of all kinds. Incorporation into the world market destroys both the autonomy of local, subsistence oriented economic production systems, and erodes local self-sufficiency, especially in the crucial agricultural sector. As Bodley suggests with regard to the commercialization of the agricultural sector in England between 1500 and 1700 A. D., for example, this extension of the commodification process implied that, "the market value of grain ultimately became more important than its nutritional value" (2001:116)[100].

Taken together, the imposition of state control, the commod-ification of land, and the commercialization of formerly self-sufficient, subsistence economies completes the proletarianization process by setting up the preconditions for economic rather than political compulsion[101]. In other words, if the entire means of pro-duction–including the land itself–is privately owned, then a land-less working class is created which has no choice but to sell their labour to the capitalist class in return for wages. Similarly, if local cultures are incorporated into states which require their citizens to

---

[100] For his complete discussion of the commercialization of the agricultural sector in England and, somewhat later, in the United States see Bodley (2001: 115-17 and 140-5).

[101] While contemporary resistance to the expansion of the capitalist system on the part of indigenous peoples shall be briefly considered below, there is also a wealth of studies of resistance to proletarianization and the capitalist system in non-tribal societies. See, for example, Ong's (1987) study of factory women, and Scott's (1985) classic study of peasant resistance, both of which focus on Malaysia.

pay taxes in cash, then people have no choice but to enter the market economy to some degree. Thus, the capitalist class need not use direct political coercion to extract surplus–arranging the ownership of production systems so that offering jobs to people who have no other choice but to work for *someone* is sufficient. In a capitalist mode of production, therefore, the social allocation of labour is accomplished largely through economic or commercial means, and social power resides primarily with those who control commerce, markets and the production system.

One way of describing the historic change from domestic, to political, to commercial scale cultures, is as a progressive break-down in "entitlements"–from *direct entitlements* to *exchange entitlements* (Bodley 2001;128-9)[102]. In domestic scale cultures everyone had an equal right to that which they required in order to survive. People were entitled to an "irreducible minimum," or had a direct right to food and other basic necessities because they were members of the group. In political scale cultures, while peasantries had no direct right to food as such (which could be expropriated by the state), peasants did retain an entitlement to land with which they could produce food. As Bodley (2001) suggests, unlike in the capitalist system, it also remained in the best interests of the landed class that all of the peasants had access to land. This is because the extraction of tribute from the peasantry was the basis of social power in political scale cultures. The more peasants who were farming the land, and the more they produced, the more tribute and wealth could be collected by the state. Thus, the internal rationale of the system did promote the maintenance of the peasants, even if in poverty in many cases, as well as the expansion of the system if possible.

---

[102] Bodley bases this distinction on the earlier work of Sen (1981).

In commercial scale cultures, on the other hand, control of capital is the basis of social power and there is a complete breakdown in direct entitlements. The centralization of ownership of the means of production leaves the vast majority of humanity with no entitlement to food, nor land, nor even to a job. After all, it is in the best interests of the capitalist class to have high rates of unemployment, since this keeps wages down. It also allows for the extraction of more surplus value, the basis of their own social power. In the capitalist system, food and other resources go to those who can pay for them—or are able to purchase an exchange entitlement—not to those who need them. One is "entitled" to—or has a *right* to—only that which one can afford.

2. THE GREEN CRITIQUE: THE QUESTION OF APPROPRIATE CULTURAL SCALE

> These last one hundred years have been the time of most difficult struggle, but they have not broken our spirit nor altered our love for this land nor our attachment and commitment to it. We have survived as a people. Our attachment means that we must also commit ourselves to help develop healthy societies for all people who live upon this land. But we will not be able to contribute unless we have the means first to develop healthy societies for ourselves (David Courchene, as cited in: McNickle 1973:147-80)[103].

> Native Americans have consistently attempted to maintain a harmonious relationship with their lands in the face of tremendous pressures to assimilate. They have expressed in multiple ways that their land and the maintenance of its ecological integrity are key to their physical and cultural survival (Cajete 2000:211).

WHILE POLITICAL ECONOMY and a cultural scale emphasis have much in common, particularly concerning their portrayal of the social consequences of the expansion of the world capitalist system for non-Western peoples, perhaps the best way to contrast them, and their implications in practice, is through a discussion of their critiques of globalization. While ecological issues are at the core of the cultural scale critique, they are seldom emphasized in the political economy literature, primarily because it remains political *economy*. In fact, the eco-holist, or political *ecology* critique is often completely dismissed, as it is by Teeple (2000:149-50)[104], and by McGrew in his discussion of three predominant discourses on globalization (2000:362)[105], each of which portray globalization

---

[103] At the time the above statement was submitted to the Indian Affairs Branch of the Canadian government (October 1971), Courchene was serving as president of the Manitoba Indian Brotherhood. The statement was entitled "Wahbung, Our Tomorrow," and was quoted in *The Montreal Star*, Oct. 16, 1971.

[104] In one paragraph!

[105] McGrew lumps the eco-holist emphasis upon appropriate scale in with the "radical school," discussed below, and makes a point of associating both with various "racist, xenophobic, and reactionary groups [who] also seek to combat the 'tide of globalization'" (2000:362).

and its consequences somewhat differently.

The pro-globalization philosophy is articulated by the "neo-liberal school" (McGrew 2000:348-50), which can be characterized as a type of "neo-modernization theory." Modernization theory was the primary philosophy which guided international economic development efforts in the post World War II era[106]. This theory generally assumed that modernism, or a Western way of life, was something which all societies should aspire to (whether they knew it or not), and that modernization should be encouraged globally. In this view, "traditional" societies, which were characterized as "backward" and "static," were contrasted with the "dynamic" and "progressive" characteristics of "modern" society.

Rostow (1960), for example, proposed a model based upon the historical development of capitalism–or modern society–which consisted of five stages. The first was "traditionalism," which was characterized as an impediment to development, and the last was a modern society, characterized as an "age of high mass consumption." In order to create an economic "take-off" similar to that of the Industrial Revolution in the West, Rostow proposed that two conditions were necessary. The first was the accumulation of capital, which happened in the age of long distance trading in the European context, and which was provided for in the modern-ization era by international development loans. The second precondition was breaking away from "traditional" values–such as loyalty to kin groups, community and the land; egalitarianism, reciprocity and common property–and their replacement with the Western values of individualism and entrepreneurship.

Essentially, modernization theory argued for the *assimilation* of non-Western peoples to Western ways of living, and for the

---

[106] For a good discussion of modernization theory see Lewellen (1992:151-4).

"progressive" homogenization of human cultures and societies into dominant Western patterns. This ethnocentric assumption is also shared by contemporary neo-liberalism, which has characterized American foreign policy since the late 1980s. As Lewellen suggests, "U. S. foreign policy has consistently been based on one form or another of modernization theory" (1992:154).

As Robbins describes it, neo-liberalism promotes the following platforms: first, that economic growth is equivalent to progress; second, the promotion of free markets, unrestrained by government regulation; third, economic globalization, or the removal of "barriers to the free flow of goods and money;" and finally, the privatization of government functions and services. The primary functions of government are seen as the provision of necessary infrastructure and the protection of private property rights (2002:89-90)[107].

Globalization is not a problem from this perspective, but rather the solution. Economic growth and the expansion of the market system are portrayed as the path to progress and prosperity for all. The wealth created by the capitalist class, we are told, will eventually (and mysteriously) "trickle down" to the impoverished masses, thus ending poverty and malnutrition and bringing social prosperity to all.

The "transformationalist school," to which McGrew himself belongs, is the position of the traditional reformist left (2000:351-2). While certainly not pro-globalization (in the sense of the unfettered expansion of capitalism), transformationalism argues that recent transformations in the forces of production have made the nation-state obsolete as a legitimate site of resistance to such developments. This is primarily due to the fact that the regulation of

---

[107] For a much more detailed consideration of neo-liberal policies and their rationale see Teeple (2000:81-131).

international capital does no longer takes place at the national level, but at the level of supranational agencies, and supranational agreements concerning free trade and investment[108]. Thus, regulation takes place at a level which is beyond that of states themselves, which is the only level of social organization which is (ostensibly) committed to democratic ideals, and to the protection of the rights and the well-being of their citizenries.

Importantly, as Teeple suggests, there is also a contradiction between the principles of Western political democracy at the state level and market values:

> The commitment to market principles and practices as embodied in neo-liberalism is antipathetic to the principles and practice of democracy. The reason is straightforward: the free market and democracy represent *in principle* two contradictory forms of resource allocation for society (2000:126-7, original emphasis).

Where the former is based upon "economic justice," the latter is based upon "distributive justice," or the social welfare institutions of the post-World War II era. Teeple characterizes the latter as a sort of "compromise" worked out between national capitals and national working classes during the Cold War era, when capital was still primarily organized nationally, and the communist option was still a realistic threat. While it is critical of the non-democratic character of economic globalization, however, and the threat which it poses to social democracy, transformationalism is not a critique of global scale organization as such. In fact, globalization is pretty much accepted as an "inevitable" result of the transformation of the

---

[108] Similarly, following the sociologist Leslie Sklair (1991), Bodley suggests that "it is no longer very helpful to focus on the activities of nation-states because the primary agents in the world system are all *transnational:* (1) transnational corporations (TNCs), (2) the transnational capitalist class (TCC), and (3) the transnational mass media and advertising" (2001:20-1, original emphasis). Compare this to Robbins' analysis, which focuses upon the relations between consumers, labourers, capitalists and nation-states (1999:7), rather than focusing upon the scale of social interactions.

means of production made possible by contemporary cybernetic and information technologies. The solution in this view is to "scale-up" resistance to match the scale of the threat, especially through creating equally global means of resistance to the capitalist class on the labour, social welfare and environmental fronts (Teeple 2000:150, 198-99), and/or global forms of democratic participation.

Finally, the "radical school" is described as the position of the more radical "left." This view characterizes globalization as "nothing less than a new mode of Western imperialism, in which multinational capital has come to replace military power as the instrument of domination" (McGrew 2000:350-1). It can be characterized as a type of "neo-dependency theory[109]," which suggests that neo-colonialism, and the spread of economic compulsion, also has an international dimension. For economic compulsion acts not only upon individuals or peoples at a local level, but also upon states at the international level. This is especially true of those states who are indebted to international financial institutions such as the World Bank and the International Monetary Fund (IMF).

The consequences for debtor nations who cannot make their scheduled debt payments are an excellent example of the manner in which such international institutions erode national sovereignty and the ability of states to create independent national policy. The usual response of international financial institutions has been to impose various types of "structural adjustment" policies to ensure debt repayment. These programs often include the imposition of such neo-liberal policies as the privatization of government property, reductions in spending on social programs, the devaluation of the county's currency, and policies which attempt to increase revenues by increasing export production of primary

---

[109] For a good discussion of dependency theory see Lewellen (1992:156-8).

resources (Robbins 1999:106).

As Bodley suggests, the type of "'export led' economic growth" promoted through structural adjustment programs "requires debtor nations to extract as much as possible from their labour force and their natural resources and provides no incentive for sustainable resource management or social equity" (Bodley 2001:210)[110]. These policies can also have devastating consequences for indigenous peoples and their traditional lands. This is because the mad scramble to harvest and export raw materials to fuel continued capitalist expansion which results often leads to further incursions into the territories of indigenous peoples, as can be seen, for example, in the Brazilian Amazon (Berwick 1992).

The objective from the "radical" perspective, according to McGrew, is therefore to protect or restore national sovereignty. This is once again portrayed as being eroded by the growing influence of transnational corporations, international financial institutions, and the growing reach of international free trade and investment agreements. For all put constraints upon the ability of states to set independent national policies which are in the interests of their own people. Thus, these developments are often portrayed as a threat to democracy, which operates at the national, rather than the international level. As Robbins suggests:

> the economic goals of capital controllers–profit, a guarantee for private property, and little risk–can often conflict with larger societal goals, such as relative economic equality, environmental safety, equal access to medical care, and equal access to food. In other words, making the world safe for capital sometimes means making it unsafe for people (1999:111).

---

[110] As Bodley suggests, international debt is also a means of economic exploitation, or of siphoning money out of debtor nations. Between 1982 and 1990, for example, $418 billion more returned to the rich creditor nations in payments of debt and interest, as was transferred from the rich nations to the poor through all forms of economic assistance, charity, private investment, and bank loans" (2001:210).

The cultural scale perspective does share many aspects of the social critique provided by the various approaches within the political economy literature. These include the critique of the spread of economic compulsion as a neo-colonial process, and of the anti-democratic character of economic globalization. They differ widely, however, in their prescriptions for future action and resistance in response to these trends. Where transformationalism suggests that we should "scale up" resistance in order to match the global scale of the threat, and the "radical" school suggests that we should "scale down" to the national level in order to protect democracy, sovereignty and the environment at a national level, scale theory suggests that we should take a step further. As Esteva and Prakash suggest, in direct opposition to the transformationalist stance in which resistance must also go global:

> the true problem of the modern age lies in the inhuman size or scale of many contemporary institutions and technologies. Instead of trying to counteract such inherently unstable and damaging global forces through government or civic controls that match their devastating scale, the time has come to reduce the size of the body politic which gives them their devastating scale (1997:287).

Scale theory, not unlike the "radical" critique, suggests that we should "scale down" human organization, but to a *local* or *bioregional* level (Tokar 1987), rather than a national one. In other words, we need to organize at a scale which is similar to that of domestic scale cultures, which tended to confine themselves to specific territories or habitats, with particular ecological characteristics. As Spretnak and Capra point out, "[b]ioregionalism has taken on a deeper meaning than mere localism, one more akin to the Native American sense of abiding respect for the natural forces and the surrounding life forms" (1986:204).

From an eco-holist, or cultural scale perspective, there are many reasons for scaling down to a local or bioregional level of organization. Not surprisingly, many of its arguments also closely mirror indigenous calls both for the restoration of local autonomy and ecological sustainability.

First, a cultural scale perspective has long been critical of centralized organization's tendency to erode democracy. In fact, a useful way of characterizing the cultural scale perspective is *as a critique of centralized planning*[111]. Ironically, during the Cold War era, Soviet communism and the capitalist system were often contrasted in both academia and the popular media using an opposition between centralized planning and free market systems. Yet it should now be obvious to anyone in a world in which many of the largest financial entities are now corporations rather than states[112], that centralized planning is just as essential to the operation of transnational capitalism as it was in the operation of the Soviet economy, if not *more* so. In hindsight, the true differences between communism and capitalism appear to be that capitalists, rather than the state, own the means of production, as well as the *scale* at which they are organized—nationally or *internationally*. Effectively, this makes the globalization of the capitalist system of corporate ownership *centralized planning on a global scale*—just as we were afraid of during the Soviet era. After all, such massive international conglomerates could not function without centralized

[111] Teeple also suggests that transnational corporate planning is increasingly characterized by "a form of 'central planning,' but for profit" (2000:182).

[112] Robbins (1999:137) provides a list of the top 100 financial entities globally—based on their gross annual production—as of 1991. Fully half at that time were corporations rather than states, with the largest corporation in the 21st position. Bodley suggests that by 1998, "revenues and sales of just 500 of the world's largest companies represented 40 percent of the global GNP" (2001:22), while Robbins points out that just the top three "ultra-rich people...own more than the gross national product of the forty-eight poorest nations combined" (2002:100). All of these statistics suggest that corporate concentration has reached alarming levels globally.

planning, and with many sectors coming to be controlled by a mere handful of corporate conglomerates, it is surely questionable how "free" the market remains in the era of transnational capitalism[113]. As Teeple relates:

> one can speak of a market only when there are numerous competing corporations for any given product, and no corporation with a dominant share of the economic activity...There is no longer any economic sector that fits this description, there is only more or less monopoly power...Here, largely unfettered by political considerations, is a tyranny unfolding–an economic regime of unaccountable rulers, a totalitarianism not of the political sphere but of the economic (2000:154)[114].

While Teeple is referring to the relatively recent growth of transnational capital, scale theorists recognized the anti-democratic character of centralized planning decades before the term "globalization" captured the popular imagination. As Halweil makes this point, "Political scientists have long recognized that a broad base of independent entrepreneurs and property owners is one of the keys to a healthy democracy" (Halweil 2001:157)[115]. In other words, the more decentralized is the pattern of ownership of the means of production, the more *democratically* organized is the *economic system*.

As Schumacher pointed out in the late 1970s, "industrial society, no matter how democratic in its political institutions, is autocratic in its methods of management" (1979:28). While the Western world

[113] As Halweil suggests, for example, "[t]hree conglomerates...dominate virtually every link in the North American (and increasingly, the global)" food system–from inputs, to corporate farms, to marketing and processing, to brand name products. He describes this hierarchical pattern of economic organization as "vertical integration," i.e. involvement in every stage of the production and distribution process, which allows them–essentially–to buy from themselves at each stage to the greatest degree possible, and to keep basic commodity prices as low as possible (2001:154).

[114] Bodley makes a similar point, suggesting that "in real markets competition tends to be self-limiting as monopoly power increases" (2001:230).

[115] Halweil is referring to the classic study of two agricultural communities by Goldschmidt (1978). See also: Bodley (2001:142-3).

proclaims very loudly its adherence to democratic ideals in the international arena, they seem to be conveniently forgotten when it comes to the organization of its own business activities, for that is not in the best interests of the capitalist class. And because of the false, but politically convenient, dichotomy between political and economic forms of organization, Western countries continue to call themselves "democratic," even while their economic organization, which is now their dominant social institution, is increasingly centralized and authoritarian. For, as Schumacher continues, "such centralized forms of rule cannot possibly preserve order without authoritarianism, no matter how great the wish for democracy might be" (1979:29) In other words, they cannot function without economic compulsion. And as Singh relates:

> It would be emphasizing the obvious to say that industrial capitalism forced the world, almost from its beginnings, into a hierarchical mould, with a mere handful of metropolises towards the top and a multitude of colonies and subcolonies underneath (1976:44).

Not only are transnational corporations hierarchical in their internal patterns of organization–the very antithesis of freedom and egalitarianism–the entire world is also forced into a hierarchical pattern of "haves" and "have nots" as well. For poverty in the Third World, as in the inner cities and remote, rural areas of America itself, is an inherent part of the functioning of the system. After all, the activity of accumulating wealth on which capitalism thrives also *implies* that it will accumulate in the hands of very few persons and, as experience also shows, in very few nations. So as Wendell Berry, another influential scale theorist has suggested, "as a social or economic goal, bigness is totalitarian: it establishes an inevitable tendency towards the *one* that will be the biggest of all" (1977:41, original emphasis). Thus, as Bodley points out, "[s]cale theory offers a liberating approach to contemporary problems because it identifies scale itself as the principle problem" (2001:226).

To summarize the critique in terms of centralization's anti-democratic character, scale theory makes two contributions. The first is its emphasis on the fact that a decentralized pattern of economic ownership is more democratic, even if it remains privately owned. After all, this implies that a greater percentage of the population have greater (and more equal) access to the means of production. Thus, not unlike Native American philosophies, eco-holism applies egalitarian ideals to both the economic and the political spheres.

Second, democracy itself is also shown to work best at a smaller scale. Participatory democracy, the most egalitar-ian pattern, appears to work well only in relatively small groups, or at the domestic or community scale[116]. Representative democracy, on the other hand, becomes less and less representative, and less respons-ive to the actual needs and desires of the people represented with increases in scale. As Winter suggests:

> [o]ur notion of 'democracy' is, in reality, a system of elite decision-making. According to the rules, the public is almost entirely reduced to the role of spectators, with allowance for a periodic ratification of the elite rulers every four or five years (1997:159).

As a social critique, scale theory is a defense of democratic principles. What it points out is that global-scale organization erodes the prerequisites for democracy, namely: (1) a relatively small scale form of social organization, and (2) relatively equal access to the means of production.

Yet the critique of central planning, or "globalization," does not stop with its anti-democratic character from a cultural scale perspective. The critique has both social *and* ecological aspects. Two

---

[116] Or perhaps even at the scale of the small, locally owned business firm for that matter...

closely related critiques of particular note are that both global-scale thinking and organization, as well as cultural homogenization on a global scale may both preclude successful adaptation.

The first of these critiques is closely related to eco-holism's critique of the reductive, non-qualitative nature of modern scientific and economic thought, which currently guides such centralized planning. Following Wendell Berry, Esteva and Prakash suggest that such "global thinking" is both arrogant and illusory. As they state:

> [w]e can only think wisely about what we actually know well. And no person, however sophisticated...can ever 'know' the Earth–except by reducing it statistically, as all modern institutions tend to do today, supported by reductionist scientists (1997:279).

In other words, as I often express this to my students, "you cannot adapt to that which you do not know," and not merely as a statistical abstraction in one's computerized data base, but through direct, first hand experience of it. Or as Cajete suggests, "to know any kind of physical landscape you have to experience it directly; that is, to truly know a place you have to live in it and be a part of its life process" (2000:181). Further, "[w]hat most people know about animals and nature comes from television," and "[w]hile moderns may have technical knowledge of nature, few have knowledge of the non-human world gained directly from personal experience" (Cajete 2000:23).

This suggests a distinction between *abstract* knowledge–of the type provided by reductive thought, or "broadcast on the television screen"–and *concrete* knowledge–based upon direct experience. Further, the local scale is the *only* scale at which "humans can really understand, know and take care of the consequences of their actions and decisions upon others" (Esteva and Prakash 1997:278-9). Not only is the local, or "human scale"

(Schumacher 1998:60) the only scale at which we can truly know anything in any concrete way, it is also only at the local level that we are able to receive immediate feedbacks on the social and ecological consequences of our actions. So as Schumacher proposed "[t]he administrators of a large organization cannot deal concretely with real-life problems and situations: they have to deal with them abstractly" (1998:60).

The problem is that the feedback provided by direct knowledge is *precisely* the type of feedback which is *required* for any sort of adaptive responses to changing circumstances. It is also necessary to inspire changes in behavior which could correct for the negative ecological consequences associated with particular choices and actions[117]. As Berkes and Folke suggest, "the key factor in successful adaptation may be the presence of appropriate feedback mechanisms which enable consequences of earlier decisions to influence the next set of decisions which make adaptation possible" (1998:19). Without concrete knowledge, such feedbacks are indirect at best, and potentially damaging at worst. As Bodley concludes:

> the increasingly hierarchical structures of more complex cultural systems tend to become maladaptive. Higher-level decision makers are likely to be inadequately aware of the local impacts of their actions (2001:27).

Bodley expands further upon this critique of increases in scale in his discussion of the manner in which centralization tends to subvert the normal negative feedbacks which often helped to prevent domestic scale cultures from adopting unsustainable patterns of resource use. In fact, he suggests that the breakdown in direct feedback caused by centralized patterns of thought and organiza-

---

[117] This also provides a further critique of centralized planning, since "growth in scale concentrates social power, [and] proportionately fewer people make more important decisions for ever larger numbers of people" (Bodley 2001:226). Such decisions also effect areas and peoples of which they have little or no direct knowledge, which is one reason why they tend to be so damaging.

tion may be one of the "roots of the environmental crisis" (2001:42-3). As he puts it, through the growth in scale of both thinking and social organization, "global commercial culture has short-circuited the normal cultural feedback mechanisms that prevent over-consumption," or the unsustainable use of resources. Bodley singles out four ways in which it has done so. These include commercial scale culture's: (1) dependence on nonrenewable resources and (2) imports; its tendency towards (3) increased urbanization, and (4) consistent increases in institutionalized inequality (2001:42-4).

The first–dependence on nonrenewable resources, such as fossil fuels–is problematic because it has led to an enormous over-consumption of resources beyond what can be sustained through relying upon renewable resources alone. The second and third–dependence on imports and urbanization–both remove us from direct feedback concerning the social and ecological con-sequences of our actions. This is because both imply a situation in which the vast majority are relying upon energy, food and other necessities which are produced far from one's local area, and about the production of which one has no direct or concrete knowledge.

Finally, increases in institutionalized inequality also subvert direct feedbacks or knowledge in at least two ways. First, on an international scale, it is regional differences in wealth which allows the relatively affluent to consume a disproportionate share of the resources of less affluent areas, concerning the impacts of which they again have no direct knowledge. Second, through in-stitutionalized inequality, it is primarily the capitalist class which controls the popular media–a highly edited and selective portrayal of reality to begin with–where corporate concentration is just as pronounced as in any other economic sector (Winter 1997). And it is they who control the editing and selection process, allowing them to centrally manage, to a large extent, the types of abstract feedbacks which the average person does receive on the consequences of their actions.

Thus, all four processes subvert the normal negative feedbacks through which concrete, local knowledge is gained. The result, as Esteva and Prakash conclude, is that "[t]he modern 'gaze' can distinguish less and less between reality and the image broadcast on the television screen" (1997:278). Or as Cajete phrases it, "[t]he 'mediated pulse,' that window through which we know nature today, is largely only a fantasy created through words and images describing nature" (2000:267). In other words, the modern gaze is increasingly unable to distinguish between abstract and concrete knowledge, as the information upon which our decisions are based becomes further and further removed from direct contact with nature. Yet "true knowing is based on experiencing nature directly" (Cajete 2000:66). What we require, as Harries-Jones suggested, is "an appropriate epistemology of survival" (1992:160), that is, a way of knowing which is appropriate to the local and bioregional scales at which successful adaptation operates.

Global-scale thinking not only relies on abstract, reductive knowledge which operates beyond the level of local-scale ecological feedbacks, it also tends to completely ignore qualitatively important characteristics of organic systems. This includes the crucial importance of local ecological and cultural diversity. Consequently, the global system has tended to apply and promote the same patterns of organization, technologies and production strategies around the globe, irregardless of local ecological and cultural diversity.

The long-term trend has been towards the erosion of both, and towards increasing cultural and ecological homogeneity. As Bookchin relates, "[t]he homogenization of ecosystems goes hand in hand with the homogenization of the social environment" (1991:138). Similarly, as Winona LaDuke, the Anishinabe author and activist suggests, "[t]here is a direct relationship between the loss of cultural diversity and the loss of biodiversity" (1999:1). For both cultural and ecological diversity have always been the common victims of the expansion of commercial scale culture, or to use Bodley's (1999) provocative title, the common "victims of progress."

The problem is that both are essential for survival, though in different ways. Ecologically, biodiversity is now widely considered to be the key to the *resilience* of ecosystems. As Bodley suggests, resilience "refers to the ability of systems to avoid collapse and retain their shape and scale in the face of various stresses or shocks" (Bodley 2001:24)[118]. In other words, the more species in an ecosystem, the more resilient it is, and the more resilience, the more likely it will be able to absorb disturbances, and the less its vulnerability to collapse. The problem is that the expansion of the capitalist system has been eroding biodiversity, and thus ecological resilience, on a global scale. So much so, in fact, that the present rate of extinction is now commonly compared to that of previous "mass extinction" events, such as that in which the dinosaurs were destroyed (Hosansky 2001; Quammen 1998)[119].

The erosion of cultural diversity on a global scale, and the replacement of locally diverse cultures with an increasingly homo-

---

[118] The importance of a focus on ecological resilience was first proposed by C. S. Holling. See, for example, Holling (1973).

[119] One of the reasons for the increasing rate of extinction is also related to the spread of global trade: the spread of "invasive species" through the global trading system into areas in which they have no natural predators, which often leads to the displacement of local species. See Bright (1999) and Hosansky (2001).

genized world culture–which is the logical outcome of current trends–is equally problematic (Gadgil 1987). As Berkes suggests, "cultural diversity is akin to biodiversity as the raw material for evolutionarily adaptive responses" (1999:79). Consequently, Bodley proposes that "[c]ultural homogeneity, or loss of cultural diversity, fostered by the commercial-scale culture could objectively be regarded as maladaptive because diversity is the basis of cultural evolution" (2001:25). In other words, cultural diversity is ncessary if human societies are to succesfully adapt themselves to local ecological diversity. The reason cultural diversity is so crucial, is that it allows for maximal flexibility in human responses to change at the local level. This leads to increased resilience, or adaptability to disturbances and changes, at the cultural level. From an eco-holist perspective, therefore, "conserving cultural diversity" is an "ethical imperative" (Berkes 1999:28).

Human cultural or technological monocultures on a global scale, on the other hand, imply a massive decrease in resilience and flexibility. As Chris Bright (2001:22) suggests, "monoculture techno-logies are brittle," not only because they reduce our options as a species, but because all of our eggs are in one basket, as it were. Thus, if our technologies or cultural strategies have negative ecological consequences, a global monoculture ensures that these negative effects will be experienced globally as well. Similarly, if resources crucial to the global system fail, unlike in the collapse of previous, regionally organized states, the social collapse will again be global. As Kohr himself suggested, therefore, local scale organization is advantageous for adaptive behaviour, "because the ripples of a pond, however animated, can never assume the scale of the huge swells passing through the united water masses of the open seas" (as cited in Esteva and Prakash 1997:287).

Taken together, these two points suggest two things. First, that the local or bioregional level may be the *only* scale at which adaptation can successfully operate. This is because it is the only level about which one can have concrete knowledge, the only level about which one can receive direct feedbacks, and consequently the only level at which one can adapt or respond to local ecological diversity in a successful manner. In fact, as Bodley suggests, "domestic-scale cultures are humanity's only cultural system with an archaeologically demonstrated record of sustained adaptive success" (2001:13). All previous centrally organized societies have had a tendency towards periodic institutional collapse, often for ecological reasons.

Second, these points suggest that--given a preexisting pattern of ecological diversity--one would expect that the only successful pattern of human adaptation would involve an equally diverse pattern of local cultural strategies and adaptations. This is especially so if we are to follow natural archetypes, and to use the diversity of local and bioregional ecosystems as a model after which to pattern the organization of human societies. In other words, if we are to fit ourselves intelligently into the patterns of natural systems, as many Native American societies seem to have attempted to do, we must primarily do so at a local or bioregional level, appropriately to the locally diverse ecological and climatic conditions which exist in those areas.

Consequently, Esteva and Prakash propose that rather than global thinking, or global scale resistance, what we need is a type of "radical pluralism." This way of thinking is "conceived for going beyond Western monoculturalism." The goal is locally focused and locally diverse forms of resistance, which form coalitions to support one another's efforts to oppose global forces, and to

protect communities and ecosystems at the local level (1997:286)[120].
And as they put it themselves:

> the solidarity of coalitions and alliances does not call for 'thinking
> globally.' In fact what is needed is exactly the opposite: people thinking
> and acting locally, while forging solidarity with other local forces that
> share this opposition to 'global thinking' and 'global forces' threatening
> local spaces (1997:281-2).

Thus, resistance must begin and end with the local. It is only at this
level that normal people can act, and "the struggle against Goliath
enemies demands that there be no deviation from local inspiration
and firmly rooted local thought" (1987:282). After all, "'global
forces' can only achieve concrete existence at some local level, and
it is only there at the grassroots that they can most effectively and
wisely be opposed" (1987:280).

A cultural scale approach, therefore, appears to be the only
critique of globalization which provides any direct support for
contemporary indigenous aspirations towards economic and
political self-determination. It essentially argues that we need to
reestablish such a pattern of local economic and political organization on a global basis in order to reestablish adaptive patterns of
life—and not just for indigenous peoples, but for everyone. Thus, it
is not only a defense of indigenous cultures, but of *rural* cultures
more generally. For centralization, the commodification of land,
and urbanization have always implied the depopulation of the
countryside, and the more general destruction of rural culture,
even in the West (Berry 1977, 1981; Halweil 2001)[121]. As Cajete concludes, "the preservation of cultural diversity, and in turn,

---

[120] Esteva and Prakash's argument is a response to the popular ecological
slogan, "think globally, act locally," which seeks to "extend the valuable
insights in the second part of the slogan to the first part" (1997:278).

[121] Of which I have direct knowledge, having been raised on a farm in southern
Manitoba, Canada, an area where I continue to reside, as of this writing,
though my career requires that I reside in the city of Winnipeg.

contemporary revitalization of cultural ecological philosophies world wide...must begin with our own communities and bioregions" (2000:267).

3. THE GREEN FROG-SKIN ILLUSION: ECOLOGY AS A CHALLENGE TO ECONOMIC HEGEMONY

> it might be said that the main function of a system of domination is to accomplish precisely this: to define what is realistic and what is not realistic and to drive certain goals and aspirations into the realm of the impossible, the realm of idle dreams, of wishful thinking. There is certainly much to be said for this limited construction of hegemony, since it recognizes the vital impact of power on the definition of what is practical (Scott 1985:326).

> Economics plays a central role in shaping the activities of the modern world...there is no other set of criteria that exercises a greater influence over the actions of individuals and groups as well as over those of governments (Schumacher 1973:40).

BESIDES ITS CRITIQUE of centralised planning, which eco-holism ex-presses through its emphasis upon appropriate cultural scale, eco-holism also provides a further critique of Western society and its ecological impacts. This is its critique of Western economic *logic*, and its consequences. Eco-holism argues that ecology represents a fundamental challenge to economic hegemony, or to the dominance of commercial values. In fact, it suggests that ecology is far more than a mere *supplement* to economic thinking. Rather, it is best seen as an empirical, or scientific *replacement* for it. Perhaps this is why they share the same Greek root,"eco," since both are designed to think about the same sorts of questions. This does not imply that it is immediately practicable to dispose of contemporary economic institutions, but rather that ecological considerations must be given priority over the economic in social policy decisions. In other words, as Rolston so succinctly put it, "the bottom line ought not to be black unless it can also be green" (1988:325).

An eco-holist perspective thus proposes a fundamental change to the criterion by which the "successfulness" of human action is judged. Indeed, it proposes that the maladaptiveness of the current practices of Western culture stem not only from the inappropriate scale of its social organization, but from an aberration in its dominant system of values as well. Quite simply, these too often places the value of profit above the more basic value of life, as shall be discussed below.

The latter is also a point which has been made, either explicitly or implicitly, in Native American literature for some time. In fact, Native thought seems to demonstrate a consistent skepticism concerning the very reality of the "economic" sphere, at least as it is traditionally conceived of in the West, and of the patterns of life and thought inspired by it. As Lame Deer suggested, with reference to what was described as "economic compulsion" above, "even a medicine man like myself has to have some money, because you force me to live in that make-believe world where I can't get along without it" (1972:37). Further, Native literature often makes the same criticisms of the anti-democratic character of economics, as well as of the preeminence of economic values, as did the various authors discussed above, for as Solomon relates:

> Governments in Canada and the USA are called democratic. Nothing is further from the truth. That is a giant delusion. The real power is somewhere else...In Canada there are about 8 families who control this country. I call them the invisible government who really call the shots...What we have is *Criminal, Captive, State Governments* who do what they are told by the elite (1994:67-8, original emphasis).

Upon contact with Western culture, indigenous peoples around the world, past and present, seem to have had little trouble identifying the central values which it pursues. And Western culture's central values are rarely identified with the religions which it *professes*, but rather with the religion which it *lives*. In the words of Harold Cardinal, a prominent Cree activist, "[i]n Canadian society, power comes from the crackle of the almighty dollar bill. Canadian society is materialistic. It is not long on humanist tendencies" (1969:144). Or as Eastman expressed this point some 50 years earlier, "When I reduce civilization to its lowest terms, it becomes a system of life based upon trade. The dollar is the measure of value" (1977:194).

Lame Deer once described economics as "the green frog-skin illusion," and suggested with regards to money that "[i]n our attitude towards it lies the biggest difference between Indians and whites" (1972:31). Like many other Native Americans, Lame Deer also had little trouble identifying economic values as the true ethos of Western culture, or in Bodley's language, had little trouble identifying the fact that commercial values were dominant in Western culture, as Solomon argued above. Not unlike eco-holists, Lame Deer also identified the materialism of the Western world, and especially the commodification of the land, as central threats to the future of life on Earth. As he put it in his own words:

I made up a new proverb: 'Indians chase the vision, white men chase the dollar.' We are lousy raw material from which to form a capitalist. We could do it easily, but then we would stop being Indians...deep down within us lingers a feeling that land, water, air, the earth and what lies beneath its surface cannot be owned as someone's private property. That belongs to everybody, and if man wants to survive, he had better come around to this Indian point of view...because there isn't much time left to think it over (1973:35).

These quotes suggest that the Native American literature has been attempting to point out that thinking about our relationship to the natural world using economic criteria is problematic for some time, and that we take economics seriously as a guide for human behavior at our peril.

Perhaps this explains one of the most fundamental traits which eco-holists and indigenous peoples share in practice: their common resistance to capitalist expansionism and its patterns of resource exploitation. In other words, they share a tendency towards resisting neo-colonialism and the destructiveness of capitalist technology, often forming coalitions in recent years as a result.

This is a point which has already been abundantly documented in the recent literature, and will only be briefly summarized here. Indigenous resistance to neo-colonial expansionism in Canada, for example, whether it be hydroelectric development (Chodkiewicz and Brown 1999; Niezen 1998,1999; Waldram 1988), or whether it be extractive industries such as oil fields (Goddard 1991), uranium mining (Goldstick 1987), or clear cut forestry (Glavin 1990) are all well documented. Similar patterns of indigenous resistance have also been documented in the United States (Gedicks 1993; LaDuke 1999), and internationally (Bodley 1999; Gedicks 2001). One of the common themes of indigenous resistance both in the North American context and internationally, is also very well summarized by LaDuke, who is herself a Native ecological activist who has been actively involved in the Green Party in the United States:

> Grass roots and land-based struggles characterize most of Native environ-mentalism. We are nations of people with distinct land areas, and our leadership and direction emerge from the land up. Our commitment and tenacity spring from our deep connection to the land (1999:4).

Eco-holism's central emphasis upon the importance of appropriate cultural scale argue that ecological resistance should adopt much the same strategy. Thus resistance to capitalist expansionism, and its detrimental effects, represents one potential point of contact between the two. Many of the works cited above describe precisely such coalitions, and their usefulness in practice to both sides. This suggests the need for the type of "radical pluralism" proposed by Esteva and Prakash, through which local actors support one another's common struggles against common global enemies in defense of local communities, local economies, future generations and our common Mother Earth, despite their cultural or epistemological differences.

To return to a theoretical level, eco-holism in particular launches its attack on economic hegemony primarily though its critique of materialism, and of the various determinisms–be they structural, technological or economic–which a materialist approach suggests. Its critique of materialism is another one of its central differences with political economy as well. As Wolf suggests with regards to Marx, "he was by no means an economic determinist," yet "[h]e was a materialist, believing in the primacy of material relationships as against the primacy of spirit" (1982:21). In other words, believing in the primacy of the material over the ideological. This implies both a determination of the latter by the former, as well as the usual dualistic metaphysic of the Western world. For as Marx put it himself:

> The mode of production in material life determines the general character of the social, political, and spiritual processes of life. It is not the consciousness of men that determines their existence, but on the contrary, their social existence determines their consciousness (as cited in Harris 1979:55).

The critique of the materialist and determinist tendencies within the political economy literature has taken many forms. As Roseberry suggests in his review of the literature, "both dependency theorists' and mode-of-production theorists' understanding of anthropological subjects in terms of capitalist processes too often slipped into a kind of functionalist reasoning," which prioritized *structural* determinism over *agency*. However, "the emphasis on structural determination was often too strongly determinative, leaving too little room for the consequent activity of anthropological subjects" (1988:170-1)[122]. Bodley made much the same point with regards to the following passage were Robbins suggests that:

> The culture of capitalism is devoted to the production and sale of commodities. For capitalists, the culture encourages the accumulation of profit, for labourers, it encourages the accumulation of wages; for consumers, it encourages the accumulation of goods. In other words, capitalism defines sets of people who, behaving according to a set of learned rules, act *as they must act* (1999:12, emphasis added).

In response Bodley suggests that this type of structural determinism "makes a disembodied capitalism the active agent and relegates capitalists, labourers, and consumers to the role of passive entities" (2001:72). It also obscures the operation of *social power*, for as Bodley continues, this view makes "everyone appear equally responsible for the problems created" by the capitalist system, when "[i]n reality, the primary human agents who created capitalism were elites with a vested interest in increasing the scale of consumption in order to disproportionately enhance their own power" (2001:72-3). In other words, it makes the problems associated with capitalist expansionism appear to be the fault of "the system," and obscures the "dominant directing role of specific human agents" (2001:72) in the operation of the global capitalism.

Similarly, the type of "technological determinism" proposed by Teeple in his description of the stages of capitalist development, and the consequent social changes which new production technologies have precipitated, makes the direction of this development appear "inevitable;" a result of the "laws" of capitalist development, or of the "natural" functioning of the capitalist system (2000:65-71). Like Robbins' structural determinism, Teeple's technological deter-

---

[122] Lewellen makes a similar point concerning both dependency theory and world systems theory. As he states: "Dependency theory in general, and world system theory in particular, has been hotly criticized...as economically deterministic, ignoring or slighting social, cultural, and political influences" (1992:159).

minism obscures the role of the capitalist class in *directing* these patterns of technological development, in *selecting* the types of technological research they will finance, and in *choosing* the types of technologies they will implement. From the latter perspective it would appear that information technology in particular arose more from the *desire* to globalize the economy, and the willingness to *invest* in the necessary technologies, than from the reverse.

What any materialist position implies is that ideology is not causally relevant to the functioning of the system, which must be described in strictly material terms. This is certainly a politically convenient position for those who wish to support the prevailing economic hegemony. Yet people who assert that ideas have only minor importance in causal matters forget that it is because humans have ideas, symbols and language that we have science and technology, and the latter, of course, is a material factor. In other words, the pattern of influence in social and ecological systems is not a linear pattern of influence from the "infra-structure," to the "structure," to the "superstructure," as Marvin Harris' cultural materialism proposed with his "principle of infra-structural determinism" (1979:55-6). Rather, the actual pattern of influence in social systems consists of much more complex "feed-backs" between the various "levels" of the system—be they the ecological, social or ideological.

In failing to look beyond the dominant materialist tradition of Western thought, and the dualistic sensibilities which it implies, the various approaches inspired by political economy leave us in a position where neither of the two most fundamentally maladaptive characteristics of commercial scale culture may be usefully critiqued. For both the scale of cultural organization, as well as the dominant economic ideology, are considered to be determined by the material, technological or structural elements of the system. Neither, then, can be seriously questioned or challenged.

Further, as Harries-Jones suggests, "[u]ntil now, anthropology has shared the same set of economistic assumptions as the rest of social science in its treatment of ecology" (1992:161). In other words, materialist approaches to ecology, such as that of Harris (1979), tend to reduce the functioning of complex organic systems to mere matter and energy. Bateson, on the other hand, was among the first to realize that if ecology confined itself to a study of flows of nutrients through biomass, or other "material" exchanges, it could easily be reduced to "nothing but" energetics, and since energy is quantifiable, to numbers–as it has been by ecologists such as Eugene Odum (1989)[123].

In contrast, the type of ecology which Bateson developed was explicitly an ecology of *mind*. This approach focused upon form and patterning in mental or organic processes–in a *communicative*, rather than an energetic system. The emphasis is upon identifying, comparing and relating *qualitatively distinct patterns of relationship at various scales*. As in the present work, the focus is upon abductive comparisons between the patterns of relationship at the ecological, social and ideological scales in various societies.

Bateson's methodology, which I have attempted to apply and expand upon, emphasized not only the importance of energy and material in the functioning of organic systems, but also the central importance of *information*. Information was defined as "a difference which makes a difference" in the functioning of organic and mental systems (Bateson 1979:105; Bateson and Bateson 1988:17). Bateson suggested that in mental process what is most important is not the exchange of energy or material, but the recognition of *difference*

[123] Harries Jones (1995:235-42) provides a useful comparison between Odum's energetic ecology and Bateson's communicative model.

$(1979:7)^{124}$. Mind is not separated from organic or material systems in this view, nor is there a dualism between mind and nature. Instead, as Harries-Jones relates, "[m]ind is immanent pattern, available to observation, in the same way that a natural historian observes any pattern of nature" (1995:76). Both mental process and information are considered to be not only empirically observable phenomena, but also essential to the functioning of complex organic systems[125].

The centrality of information is particularly crucial in human societies. Consequently, from an eco-holist perspective, people who suggest that ideas have only a minor role in causal matters also forget that, while money is but a cultural fiction, or a social construction which has no value unless we all believe that it does, it is also an *information system* which is particularly central to human organization in commercial scale culture. Indeed, the hegemonic economic ideology promoted through national educational in-stitutions, national governments, and the corporate media, is *essential* to the functioning of the capitalist system. Further, to the extent that we all *believe* in it–to the extent that we *equate* it with "reality"–this shared *ideology* largely determines capitalism's pat-terns of development.

---

[124] For his most detailed discussion of the "criteria of mental process," see Bateson (1979:97-137). Bateson alternatively described them as "criteria of life" (1979:137).

[125] While ecologists have been slow to adopt Bateson's emphasis upon the centrality of mental process and information in organic systems, his views are beginning to gain more currency in the more recent literature. Kates suggests that "the addition of information to energy and matter...complet[es] the triad of the biophysical and ecological basics that support life" (2001:46). Con-temporary complexity theory has also begun to recognize its importance. Norman Packard, for example, has suggested that the study of complexity is "concerned with information processing thoughout the entire biosphere," since "information processing is central to the way the biosphere evolves and operates" (as cited in Lewin 1992:170).

Indeed, it is the logic of maximization–an ideology which is consistent with the desire of the capitalist class to ceaselessly accumulate wealth and social power–which leads to the expansion of the capitalist system. As Singh relates, "maximization is taken to be the very quintessence of economic behavior" (1976:115). And while such expansion is essentially only in the interests of the capitalist class, economic "growth" is also commonly promoted by the corporate media as an indication of *social* "progress." For as Bodley suggests, "[i]n the global culture...economic growth is universally recognized as the highest priority for government policy, even when what is good for the economy conflicts with the interests of particular human groups" (1999:6).

Bateson, for one, challenged the dominant economic or commercial logic with his contrast between the "ethics of optima and the ethics of maxima." As he suggested, they are "totally different ethical systems" (as cited in Berman 1984:255). This contrast was intended to suggest that if adaption to the types of complex organic systems[126] of which we are a part is our goal, we can only successfully do so based upon an adequate understanding of the nature of those systems, and of their patterns of operation. In other words, in developing his critique of the dominant economic ethic of maximization, Bateson turned to natural archetypes. More specifically, Bateson suggested, "there are no monotone values in biology," by which he meant the following:

> A monotone value is one that either only increases or only decreases...Desired substances, things, patterns, or sequences that are in some sense 'good' for the organism–items of diet, conditions of life, temperature, entertainment, sex, and so forth–are never such that more of the something

---

[126] Complexity theory in the biological sciences often uses alternative terms, such as "complex adaptive systems" (Sole and Levin 2002) or "complex dynamic systems" (Lewin 1992). I prefer the term "organic" because it is more suggestive of an organic world view. Lewin (1992) provides a good, introduction to the implications of complexity theory.

is always better than less of the something. Rather, for all objects and experiences there is a quantity that has an optimum value. Above that quantity, the variable becomes toxic. To fall below that value is to be deprived (1979:56).

In complex organic systems, values are never *maximized*, rather they are *optimized*. As diabetics know all too well, for example, blood sugar has an optimum range in the human body. Both too much and too little are damaging to human health. Hence, blood sugar must be maintained within its optimum range to prevent these negative effects. Where food or sugar are used to increase blood sugar, insulin is used to decrease it, in a constant balancing act which attempts to avoid the damage caused by either of the extremes. Similarly, too much food leads to obesity, while too little leads to malnutrition–both of which are unhealthy–and too much or too little water leads to drowning or dehydration, respectively–both of which can cause our deaths.

What applies at the level of the individual organism also applies to organic systems at larger scales. If there were too little oxygen in the atmosphere, for example, all animal life would die, while if there were too much, fires could never be extinguished. Similarly, if there is too much carbon dioxide in the atmosphere, as there is at present, it leads to global warming and all of its unpredictable effects[127].

What receives less publicity in the corporate media is the fact that if there were too *little* carbon dioxide, the global average temperature would drop sharply below the conditions necessary to maintain contemporary life. Both oxygen and carbon dioxide have an *optimum range* within the biosphere, with this optimum being maintained, under normal circumstances, through the operation of

[127] For discussions of global warming see Calvin (2000), Easterbrook (1999), Karl and Trenberth (1999), and Lashof (2000).

the carbon cycle. Where animals breath in oxygen and exhale carbon dioxide, plants take in carbon dioxide and give off oxygen–a pattern of *complementarity* which leads to an oscillation about the optimum range for both gases under normal circumstances.

This sort of oscillation in complex organic systems is also typical at the ecosystem level. The classic example from population dynamics is that of the fluctuations in population of predator and prey–with increases in prey leading to an eventual increase in predators, and the increases in predators leading to a consequent decrease in prey (and thus in predators in time). Once again this produces a pattern of oscillation around an optimum value or, in this case, an optimum population for each species.

What Bateson suggests with his contrast between the ethics of optima and maxima is that organic systems are deviation *inhibiting* systems, which operate in terms of "negative feedbacks." The classic mechanical example of a negative feedback system is the relationship between a thermostat and a furnace, where the user is able to set the optimum value, or the desired temperature. Consequently, when the temperature of the room exceeds this optimum value, the furnace will be shut off, allowing the temperature of the room to drop, while when it falls below the optimum value the furnace is turned on. This leads to an oscillation about the optimum value which is similar to the patterns of relationship characterizing complex organic systems.

The ethics of maxima of capitalist economics, on the other hand, is consistent not with negative feedback systems, but rather with "positive feedback" systems. These tend to be deviation *amplifying* systems. In other words, positive feedback systems tend to move consistently in the direction of one of the extremes, with an intensification leading to further intensification, or a decrease leading to a further decrease in some variable. The classic mechanical example of positive feedback is the breakdown in the intended, or optimum relationship between a microphone, amplifier and speaker system. For if the sound coming out of the speakers feeds back into the microphone, it creates a "vicious circle" through which the signal is continually amplified each time it comes back around the cycle. This leads to the familiar, high pitched squeal of "feedback" with which we are all familiar in popular language, and to the eventual collapse of the system[128].

Where negative feedback systems tend to *maintain* themselves by oscillating about an optimum value, positive feedback systems tend to escalate out of control, leading to their own *destruction*. A thermostat-furnace system which went into positive feedback mode, for example, would be unable to compensate for increases in temperature by shutting the furnace off. Instead, if the room became hotter the message would be that the room needed to be made *even hotter still*. And eventually, the house would burn down.

By abduction, this is *precisely* the pattern which the ethics of maxima of contemporary economics promotes, and as Bateson concludes, "we may find that money, too becomes toxic beyond a certain point. In any case, the philosophy of money, the set of presuppositions by which money is supposedly better and better the more you have of it, is totally antibiological" (1979:56). I

What economic logic promotes a positive feedback system of endless "growth," which is not only inherently unstable and destructive, but also moves according to a pattern which is totally

---

[128] The opposite pattern of positive feedback is illustrated by the relationship between a camera and screen, where the image being recording includes a playback of that image. This leads to a diminishing pattern of images within images which eventually become too small to register on the pixels of the screen displaying the image, thus resulting in its disappearance.

at odds with the successful functioning of organic systems. Thus, one of the most fundamental changes which follows from considering *life*, rather than *profitability* as the basis of value, and from following natural archetypes rather than social ones, is a rejection of the ethic of *maximization* in favour of an ethic which considers balance or *optimization* to be the ideal. And this is, in light of the above, almost a perfect analogue for Native American philosophies, and for the "principle of sufficiency" which self-sufficient, domestic scale cultures tend to practice[129].

After all, if we wish to fit ourselves intelligently into the patterns of nature, our own patterns of organization must follow similar patterns to those of natural systems. In other words, we should be seeking an optimum human population, and an optimum level of production and consumption, which meets human needs without eroding the possibility of our own survival, or the well-being of future generations. And from this perspective we can no longer *afford* to maximize either the goods of culture at the expense of nature in the ecological sphere, nor the good of the few at the expense of the many in the social sphere.

Besides eco-holism's critique of economic *logic*, an even more fundamental critique is that economics can tell us *absolutely nothing about the non-human world*. In other words, while economics cloaks itself in the trappings of modern science—especially with its purely quantitative and reductive methodology—it is *not a science of any*

---

[129] For this reason the eco-holist critique of Western economic logic is usefully characterized as having grown out of the earlier substantivist-formalist debate in the social science literature, which was essentially a debate over the universality of that logic or, to use the language of the present argument, over whether the ethics of maxima was equivalent to human nature, in which case using Western categories to describe non-Western economic systems was not problematic. For an early discussion of the debate from the substantivist position, which argued that non-Western economic systems had their own distinct logics, see Sahlins (1972:xi-xiv).

kind in its own right. For while modern science provides the dominant *world view* of commercial scale culture, economics provides the dominant *ethos*. The latter is also typically maximized at the former's expense in modern society. For as Marx so famously stated, "*religion* is an opiate for the masses," and modern economics is the *de facto* religion of the capitalist age.

This is also a reflection of the dualistic cosmology and anthropocentrism of dominant Western philosophy in practice, and of the manner through which it is institutionalized in commercial scale culture. As Wolf proposes, for example, "economics is not about the real world at all. It is an abstract model of the workings out of subjective individual choices in relation to one another" (1982:10).

In other words, to use modernism's own metaphysical criteria against itself, economics is not a science, since it deals entirely with "subjective" human values, rather than with "objective" facts about the nature of the world. Or to put it in non-dualistic language, economics is totally anthropocentric, because it deals exclusively with human values or preferences, while ignoring those of all other living organisms. The preferences and valuings of whales and elk, after all, are not encompassed by the economic calculus.

The problem is that, from a truly scientific or empirical perspective, *nature does not care what we prefer*. Nor do our preferences alter any of the negative consequences to which maladaptive behaviors inevitably lead. A discipline devoted entirely to the measurement of human preferences in a socially constructed "market," then, can tell us *absolutely nothing* about adaptive patterns of life. Bateson himself often made this point by suggesting that the phrase "God is not mocked" also applied to "the relationships between man and his ecology," for as he continued:

It is no use to plead that a particular sin of pollution or exploitation was only a little one or that it was unintended or that it was committed with the best intentions. Or that 'If I didn't, someone else would have.' The processes of ecology are not mocked (1972:504)[130].

Or as Harries-Jones puts it, "There is no question of human value making any difference when nature dies" (1992:160). Thus, as Daly and Cobb conclude with respect to economic ideology, "[j]ust as policies derived from a discipline that knows nothing of human communities are destructive of that community, so policies derived from a discipline that knows nothing of the physical world are destructive of that world" (1989:100).

The problem is that even though economics can tell us nothing about the non-human world, commercial ideology is none-the-less the dominant ideology of Western society. And in commercial scale culture we increasingly *do* allow a completely nonscientific, quasi-religious ideology to determine not only social and political policy, but also the direction of technological change. Indeed, we have increasingly come to equate economics *with* reality itself. As Schumacher suggests, for example, "[t]he religion of economics has its own ethics, and the First Commandment is to behave economically" (1973:45). The "economical" is equated with the "feasible," while the "uneconomical" is equated with the "unrealistic," or even the "impossible." As Schumacher made this point:

> Call a thing immoral or ugly, soul-destroying or a degradation of man, a peril to the peace of the world or to the well-being of future generations; as long as you have not shown it to be 'uneconomic' you have not really questioned its right to exist, grow, and prosper (1973:41-2).

Perhaps the best illustration of the power of this type of economic hegemony is the fact that economic values routinely override the empirical data provided by science in practice. In other words, if science says one thing, while economics says another,

---

[130] See also: Harries-Jones (1992:169).

commercial scale culture almost invariably does what economics dictates at the political level, or allows scientific prescriptions to be watered down in the face of economic considerations. The Kyoto Protocol on climate change, for example, calls for a reduction of greenhouse gas emissions to 1990 levels, or a reduction of about 5.2%. According to the International Panel on Climate Change, however, "the official scientific body that advised the Conference," the reduction necessary to *actually* stabilize atmospheric carbon dioxide any time in the near future is on the order of "60 to 80 percent below 1990 levels" (Flavin 2000:177).

Similarly, while science tells us that Persistent Organic Pollutants (POPs)–or synthetic chemicals which are long lasting, organic, toxic and fat soluble–are subject to biological amplification, and will thus accumulate in human tissues, such compounds continue to be marketed. And this despite the fact that many are known to be carcinogenic or mutagenic in humans, to say nothing of the possible synergistic effects which may result from being exposed to a combination of them (McGinn 2001:204-5). Yet this does not stand in the way of their being manufactured and sold for lucrative profits. Instead, as Rachel Carson pointed out long ago, governments rush to protect their citizens from the greed of the capitalist class by instituting "tolerance margins," and as she suggested:

> to establish tolerances is to authorize the contamination of public food supplies with poisonous chemicals...then to penalize the consumer by taxing him to maintain a policing agency to make sure that he shall not get a lethal dose (1962:183).

In fact, while some of the original chlorinated hydrocarbon insecticides to which Carson referred–such as DDT–have now been banned in the industrial nations, many continue to be sold in Third World countries where government regulations are more lax. Thus even such "bans" merely mitigate local effects in one part of the world, while exporting the problem to others. And since we continue to import food from such countries, we have not really escaped from the problem at all (Bodley 2001:9).

Consequently, it would appear that pollution is not merely a *byproduct* of capitalist production, nor merely a "negative externality" of capital accumulation. Rather, the term "pollution" is a rather apt description for many of the commodities the capitalist system produces as well. As Singh observes, "the real problem is not with effluent alone, but with the nature of many a commodity produced" (1976:19).

Thus, while the capitalist class *uses* science, to be sure, it does not *follow* it. Rather, economic hegemony–or the equation of what is economical with what is realistic–is used to distort science to its own purposes, to subvert those technologies which threaten their own bottom line, and to promote those which augment it. There is perhaps no better illustration of this than our continued reliance upon fossil fuels, when all of the necessary alternative energy technologies–which are both clean and sustainable–already exist. Indeed, many of the key alternative energy technologies were developed by NASA, and power almost the entire global space program[131]. Yet we continue to rely upon coal and oil even when science tells us that nonrenewable sources of energy are inherently unsustainable (to say nothing of the many negative ecological effects which attend their use), and when we know that oil in particular will be largely depleted before the end of the present

---

[131] Both photovoltaics, which convert sunlight to electricity (Quinn 2000) as well as hydrogen fuel and hydrogen fuel-cell technology (Dunn 2001), were developed by NASA, and have been in continuous use in the space program for decades, thus making their feasibility impossible to question from anything but an "economical" perspective.

century[132].

Thus, our continued reliance upon fossil fuels has absolutely *nothing* to do with the feasibility of zero-emission alternative energy technologies, nor with the superiority of fossil fuel technologies on empirical grounds. Rather, we continue to rely upon fossil fuels due to a combination of: (1) political opposition on the part of vested interests in the fossil fuel industry, which oppose the transition, and which have even actively bought out patents or companies in order to delay its implementation, and (2) a lack of political will on the part of society at large (Rosentreter 2001; Weltman 2001). For we could have begun to implement these technologies long ago if we were willing to pay the price—either through taxes on fossil fuels, or subsidies to alternative energy, or both[133]. Alternatively, we could also resort to the simple expedient of *eliminating current subsidies* to fossil fuels[134]. So as Wackernagel and Rees point out:

> In today's materialistic, growth-bound world, the politically acceptable is ecologically disastrous while the ecologically necessary is politically impossible. Developing sustainability strategies that are consistent with the ecological bottom-line therefore depends on the convergence of ecological and political practicality (1996:40).

[132] Campbell and LaHerrere (1998), for example, use the same methodology which was successfully used to predict the peak of oil production in the lower 48 American states, to predict the peak of world oil production. They propose that world oil production will peak and begin to decline around 2008.

[133] As of this writing, for example, Germany is currently leading the way in subsidizing the installation of wind turbines (Flavin 2001), while the Japanese have a national rooftop program which subsidizes the installation of PV arrays to the tune of 50% of the initial cost (Quinn 2000). Iceland has become the first country to propose a 30 year national plan to convert their transportation sector to hydrogen fuel-cell technology, whose only effluent is water vapour (Dunn 2001).

[134] As Dunn (1999:18) suggests, merely eliminating subsidies to coal in various European nations has lead to a decline in its use by up to one half. For a discussion of the billions of dollars in subsidies which the U. S. continues to provide annually to support the petroleum industry and to protect the supply see Robbins (2002:127).

The first step is overcoming economic hegemony, and the continuing equation of the feasible with the economical. What eco-holism proposes, therefore, is that from a scientific or empirical perspective, what is "economical" is quite simply *irrelevant* when it comes to deciding what is feasible. Rather, if we have the necessary knowledge, technology, energy and materials–and most import-antly, *if we have the political will*–we may develop and implement whichever technologies we so choose. Assuming, of course, that we do not allow the economic ideology promoted by the capitalist class and their media to make these choices for us, as we too often do at present[135].

What eco-holism argues is that from an empirical perspective, feasibility must be judged according to different criteria than those used by economics: specifically, (1) according to an activity's sustainability over the long-term, and (2) according to its present and future ecological and social impacts[136]. One of the classic early studies to emphasize sustainability as a critique of commercial scale culture, or "to examine the implications of the instability" of that culture (Bodley 2001:38), was Meadows et al.'s (1972) *The Limits to Growth*. This work examined the relationships between population growth, the intensification of food production, resource depletion, industrial expansion and the impacts of pollution. Using a comput-er model to project how these variables might interact over time, and by manipulating the variables to simulate various techno-

---

[135] During the Cold War era, for example, it was feasible for the United States to send men to the moon and return them safely to Earth–not because it was economical, for it surely wasn't–but because the desire to beat the Russians to the moon provided the political will to spend the money necessary to accomplish the task at hand. Obviously, then, it is still feasible today, though we now lack the political will to do so.

[136] Another concept used to make this point is that of "ecological footprints," see Wackernagel and Rees (1996).

logical "solutions" being applied to the various sectors (perfect birth control, a doubling of available resources, pollution control, etc.), they could find no way to avoid the collapse of the system before 2100 A. D. short of "stabiliz[ing] population and industrial production as quickly as possible" (Bodley 2001:40).

Thus, not only an understanding of the impacts of population, but also of levels of consumption and the relative ecological impacts of various types of technologies are all crucial to an understanding of the changes required to design a sustainable society. In fact, where the earlier literature often emphasized the centrality of population growth concerning issues of sustainability, the current literature argues that "consumption is more threatening" (Kates 2001:49).

What eco-holism suggests, then, is a set of non-economic–and *empirical*–criteria with which to judge the "successfulness" of human action in the world, which is normally judged in terms of profitability at present. This may explain why there was a 15 year gap between the introduction of the concept of sustainability and its initial entry into the arena of global policy discussions. The first United Nations' publication to emphasize the importance of sustainability was the World Commission on Environment and Developments (WCED) report, *Our Common Future*. It defined "sustainable development" as "development which meets the needs of the present without compromising the ability of future generations to meet their own needs" (WCED 1987:43). The continuing dominance of economic ideology over science may also explain why both the WCED and other national and international reports which followed consistently concluded that "further economic growth would be needed to reduce environmental deterioration and poverty" (Bodley 2001:42).

"Development," therefore, continued to be equated with economic growth. *Further* economic growth was also proposed as a solution to the problems *caused* by economic growth. This is much like a doctor saying to a cancer patient, "don't worry, we've identified the problem, it's a tumour in your lung, but as long as it continues to grow you should be just fine." Consequently, while some scholars propose that "everyone can agree that sustainable development is a good thing...there is disagreement about the ways and means of achieving it" (Turton 2002:66), eco-holism finds the very concept questionable. As Harries-Jones observes: "[m]any have suggested that sustainable development is an oxymoron and that its prescriptions lead to paradox" (1992:160). While sustainability remains a central ecological goal, it is not "development" which needs to be sustained from an eco-holist perspective–particularly if it is equated with economic expansionism, as it so often has been in the past. Rather, it is local ecosystems, cultures, and production systems which need to be sustained[137].

In contrast to political economy, then, eco-holism suggests that the fundamental contradiction in the capitalist system is not that between labour and capital (after all, labour can be bought off). Rather the contradiction is between the capitalist logic of expansion and the limits which nature imposes upon us[138]. As Singh pointed out some time ago:

> The evident incongruity between the sheer growth which it requires for nothing less than its very survival and its narrowing ecological base

[137] Both Daly's (1995:186) discussion of "money fetishism" and Wackernagel and Rees' (1996:37) contrast between "weak" and "strong" sustainability also suggest the predominance of economic over scientific ideology concerning the issue of sustainability.

[138] While some of the recent political economy literature does acknowledge both "the conflict between labour and capital, and capital and nature" (Teeple 2000:39), a fully developed discussion of the implications of the latter contradiction is seldom forthcoming.

appears...as an absolute and unsolvable contradiction of contemporary capitalism (Singh 1976:43).

What brings the critique of cultural scale and the critique of economic logic together is a focus upon technology. Schumacher, for example, argued that it was the increasing scale of production systems themselves which lead to the "four unecological trends" which he identified in the development of capitalist production systems: giantism, increased capital cost, increased complexity, and increasing violence (1979:51-3). These trends are related to the manner in which capitalist economics understands *efficiency*; for larger, more expensive and more complex production systems invariably lead to an "economy of scale." After all, the production systems promoted by the capitalist class have always sought technologies which can accomplish three tasks. First, consistent with centralized planning, they must be able to be effectively controlled in a centralized manner. Second, consistent with the logic of maximization, they must be able to produce profits in ever increasing amounts. Third, following from the last point, and consistent with economic understandings of efficiency, they must be able to minimize the costs of human labour—or the necessity of it—to the greatest degree possible. In other words, because economics defines efficiency in terms of saving time and labour (and therefore money), it invariably seeks more centralized patterns of production which progressively eliminate the need for human labour through mechanization, thus reducing costs and increasing profits.

Consequently, economic thinking has no interest in simple production technologies which everyone can comprehend, build, fix and even own for themselves, for where is the profit in that? Instead, capitalist economics has tended to promote the development of massive technologies which allow for production–and profit–to be centralized in single locations. These highly complex and capital intensive technologies subsequently displace the simpler, cheaper, less ecologically destructive alternatives. In other words, the so-called "inefficient" small producers are put out of business as both production and ownership are centralized in the ongoing process of capitalist expansion.

As with commodification and commercialization, therefore, as production technologies are priced out of reach of the average individual, they are again left with no option but to sell their labour to the capitalist class. And increasingly, it is only the rich who can afford all of the machinery which is required in order to produce products "competitively." Many "high tech" items also do not lend themselves to small-scale production.

Yet according to Schumacher, it is partly the very scale and complexity of modern production technologies which makes them so incredibly violent in their interventions into nature. For while the Earth may more easily recover from many small-scale interventions in many different locations, the sheer size, centralization and power of modern technology is often more than local ecosystems can withstand. Further, modern industry does not adapt its technologies to local ecological and cultural conditions. Quite the reverse. Capitalism has always adapted local ecological and cultural conditions to suit the needs of capital and technology, which has often involved their destruction. Indeed, as Rappaport so perceptively observed:

> The ultimate consequence is not merely that the short-run interests of a few powerful men or institutions come to prevail, but that the 'interests' of machines–which even powerful men serve–become dominant. Needless to say, the interests of machines and organisms do not coincide (1979:164).

From an eco-holist perspective, therefore, we must not only

redefine what we mean by "success," but what we mean by technological "progress" and "efficiency" as well. Progress, for example, is not "inevitable," nor is every change a change for the better, as the concept often appears to imply in the corporate media. Rather, unlike the concept of evolution, which merely implies *change*, the very idea of progress *implies* that there is a *goal* which one is "progressing" towards, or a set of *ideals* which one would like to reach. As was discussed above, Western civilization is currently "progressing" towards an increasingly centralized and authoritarian pattern of economic control, and to an ever increasing destruction of, and pollution of nature. In other words, the "goals" of Western society appear to be the "progressive" expansion of unaccountable corporate government, and the "progressive" conversion of the Earth's biomass into humans, their domesticated animals and plants, and garbage[139]. Or as the following quotation from the *Journal of Retailing*, which was published shortly after World War II, made this point:

> Our enormously productive economy...demands that we make consumption our way of life, that we convert the buying and use of goods into rituals, that we seek our spiritual satisfactions, our ego satisfactions, in consumption...We need things consumed, burned up, worn out, replaced and discarded at an ever increasing rate (as cited in Singh 1976:56).

Thus, from an economic perspective, not only population growth, but overconsumption are equated with "progress." After all, both lead to expanding profits and increases in GNP. Nor is this pattern of development accidental, for it is a logical consequence of making maximization, centralization, economic "efficiency" and the domination of nature one's ideals. It is, in other words, "busy-ness"

---

[139] Or as Bodley suggests, tongue firmly in cheek, "[i]f we insist on considering the global proliferation of the high-consumption culture to be an indication of adaptive achievement, then...[c]ulture evolves as the global biomass becomes increasingly converted to the human sector" (2001:29).

for its own sacred sake, irrespective of its social and ecological consequences, and the ends which it serves. Again, this illustrates that economic "growth" has, in effect, become a quasi-religious dogma in modern Western culture. As Cajete points out, therefore, due to the dominance of economic values over science:

> In spite of mounting evidence that the cosmology of modernism is not sustainable, we continue to be bombarded by messages from institutions and the media that somehow everything is going to be fine: just keep on supporting your governments, businesses, and educational institutions, keep consuming, and everything will be fine (2000:280).

From an ecological perspective, on the other hand, the goal is not maximization, but the development of "a style of coinhabitation that involves the knowledgeable, respectful, and restrained use of nature" (Rodman 1986:166). And only those activities which actually help us to reach this goal may be considered "progressive." Just as a commercial scale culture has developed a technology which is consistent with the ends which it pursues, so also must an ecological society. For as Schumacher suggested:

> Ever-bigger machines, entailing ever-bigger concentrations of economic power and exerting ever-greater violence against the environment, do not represent progress: they are a denial of wisdom. Wisdom demands a new orientation of science and technology towards the organic, the gentle, the non-violent, the elegant and the beautiful (1973:34).

The best outline of what such a technology might look like remains the works of Schumacher, who proposed the development of what he called an "intermediate" or *appropriate* technology. In other words, a technology which was not only ecologically appropriate, but also consistent with local, or human scaled patterns of organization and ownership. Following Gandhi, Schumacher proposed a distinction between the type of *mass production* practiced by modern Western society, and increasingly imposed upon the world through economic globalization, and *production by the masses* (1973:153-4; 1998:54). Where mass production was characterized as inherently violent, ecologically destructive, and self-defeating in the sense that it relies primarily upon nonrenewable resources, production by the masses was characterized as being more concerned with creating jobs, providing for local needs, and creating local and bioregional self-sufficiency.

In other words, appropriate technology would make use of the best of modern knowledge and experience, but would apply this knowledge to develop a technology midway between "primitive" tools and "high" technology[140]. Such an appropriate technology should also be conducive to the decentralization of the organization of production, compatible with the laws of ecology, gentle in its use of scarce resources, and designed to serve human persons and communities, rather than to make them the servants of profit and machinery (1973:154).

Schumacher suggested four propositions to guide the development of appropriate technology, or the development of sustainable production systems. First, workplaces should be created in places where people actually live, or more specifically, in *rural* areas, rather than concentrated in urban centres. In other words, in direct opposition to current trends towards centralization, production systems should be *decentralized*. Second, workplaces should be cheap enough that they can be created in large numbers. They should *not be capital intensive*, so that only elites can afford to own the means of production. Third, production methods and technologies should be *relatively simple*, so that demands for specialized knowledge are minimized, not only in the production process, but

---

[140] Schumacher contrasted the three by proposing that if high technology is a $1000 (or pound) technology, and primitive technology is a $1 technology, then appropriate technology would be a $100 technology (1973:179-81; 1998:135-6).

in terms of raw material supply, organization, financing and marketing as well. Finally, "production should be mainly *from local materials* and mainly *for local use*" (1973:175-6, emphasis added).

The central reason for these changes is the fact that ecology redefines economic understandings of *efficiency*–which focus on production processes which efficiently *maximize profits* over the *short-term*–to a focus on production processes which efficiently *optimize productivity* over the *long-term*. In other words, eco-holism proposes a more *empirical* definition of "efficiency," which focuses upon "the efficiency of energy and material usage" (Kates 2001:48). If one is able to accomplish the same purposes while using less energy and materials, then one has increased the efficiency of the production and distribution process. And as Bodley points out, "[d]ecentralization, self-sufficiency, and self-regulation would greatly reduce the need for transportation–as well as many obvious costs of urbanization and massive centralized political systems" (2001:238).

Local production systems oriented towards serving local markets also tend to be more labour intensive, and less mechanized, thus saving further energy and materials, while also creating more jobs. The global production system, on the other hand, is "objectively unsustainable to the extent that it is dependent on the petroleum subsidies...that fuel the diesel tankers and cargo jets that move long-distance trade" (Bodley 2001:232).

In sum, then, eco-holism is neither anti-scientific, nor anti-technological. On the contrary, while eco-holism argues that science must be guided by a new world view, which can encompass an understanding of complex organic systems, it actually argues that we need to be *more* scientific in the design of our production and distribution systems, not less. Similarly, eco-holism does not argue against technology as such–since the ability to develop technology is one of the things which makes us distinctly human–but rather *for* the development of clean and sustainable technologies, such as the alternative energy technologies developed by NASA.

Indeed, it is the capitalist class, with their promotion of fossil fuel technologies invented in the early 1800s and 1900s, and in their defense of the "economical" or "profitable," which are inhibiting true progress in the energy sector, especially by opposing our efforts to bring clean and sustainable space-aged energy technologies such as hydrogen fuel-cells and PV arrays down to Earth, where they belong[141].

Thus, eco-holism's dual critiques of centralized planning, and of the hegemonic economic ideology through which it is promoted, and through which anything else is dismissed as "unrealistic," also point out the manner in which this ideology has not only perverted technology into monstrous and destructive forms, but have also reduced science itself to little more than a servant of the capitalist elite and its interests in many practical respects. The empirical prescriptions of science are generally only followed when they are consistent with centralization, the maximization of profit over the short-term and capitalist expansionism, and ignored otherwise.

By redefining "success" and "progress" in terms of long-term sustainability rather than short-term profitability, "efficiency" in terms of saving energy and materials rather than time, labour and money, and "feasibility" in empirical rather than economic terms, eco-holism suggests alternative objectives for the future development of human societies–*optimization* and *sustainability*, rather than *maximization* and *expansion*. For as Meadows et al. (1992) put it in

---

[141] For a discussion of the alternative energy program being developed by the Hopi see LaDuke (1999:187-93).

their more recent work, the goal should be "sustainability, sufficiency, equity and efficiency" (as cited in Bodley 2001:40). These are goals which are far more consistent with the patterns of organization which have always characterized domestic scale cultures, including those of Native America.

Indeed, as mentioned above, eco-holism's central rejection of both objectivism and the fact/value dualism have often been echoed in the Native American literature, and continue to be expressed by aboriginal peoples in the present day. Indeed, Oren Lyons, then an Onondaga Chief and Faithkeeper, encapsulated one of the most central points of eco-holism's critique of economic hegemony quite eloquently when he stated that:

> When we talk about economics and we talk about development and we talk about money, we have to balance that with reality. We have to balance that with quality of life, with peace, with community. I think if there is anything that Indigenous people have to offer...it is that perspective. And it is fundamentally important for survival. Every question that is political is also moral. Every question...That brings responsibility to governance and governors and people (RCAP 1997)[142].

---

[142] This quote appeared in V.4, Part III: "Elder's Perpectives," Chapter 1. "Who Are the Elders?" The Onondaga are one of the member nations of the Iroquois Confederacy. Lyons words were recorded during a statement to the RCAP in Akwesasne, Ontario, on May 3, 1993.

4. LOCAL KNOWLEDGE AND ADAPTIVE RESOURCE MANAGEMENT

> other attitudes and premises–other systems of human 'values'–have governed man's relation to his environment and his fellow man in other civilizations and at other times...In other words, our way is not the only possible human way. It is conceivably changeable (Bateson 1972:493).

> Native peoples traditionally lived a kind of communal environmental ethics that stemmed from the broadest sense of kinship with all life. The underlying aim of the science of ecology, therefore, the understanding of the web of relationships within the 'household' of Nature, is not modern science's sole property (Cajete 2000:95).

AS WAS DISCUSSED above, ecological and Native activism have begun to form coalitions on a variety of issues in recent years. To this point such coalitions have largely been concerned with the ecological effects of capitalist development and technology, and with the importance of local economic and political autonomy. Consistent with the fact that successful adaptation appears to operate best at a local or bioregional level, however, and upon the importance of direct knowledge, an eco-holist approach has also increasingly come to emphasize the importance of *local knowledge* of the land and resources. Such an emphasis is again consistent with the promotion of "radical pluralism," through which local actors; be they Western or indigenous, seek to come together to combat common "global" enemies, even while recognizing cultural and epistemological differences between their respective positions or goals. In fact, one of the best illustrations of the increasing influence of an appropriate scale approach is the increasing emphasis upon participatory research and local knowledge in recent years in both the social science and the development literature. As Sillitoe suggests concerning the development literature:

> It is increasingly acknowledged outside anthropology that other people have their own effective science and resource use practices and that to

assist them we need to understand something about their knowledge and management systems (1998:223).

Because of the recent turn towards "bottom up" research, Sillitoe claims that there has been a "sea change" in the paradigms structuring concepts of development, which suggests the importance of an Indigenous Knowledge (IK) approach. Such an approach attempts to "introduce a locally informed perspective into development" (1998:224). It also attempts to make explicit connections between local understandings and practices and those of researchers and development workers.

The term "indigenous knowledge" has been used to suggest a central emphasis upon the local knowledge of indigenous peoples, or the unique, local knowledge of particular cultural groups (Warren et al. 1995). Popularized within the development literature more generally, the term "indigenous" is meant to emphasize the culture of the original inhabitants of an area, as opposed to that of global capitalism. The term "knowledge" is meant to focus attention upon the contrast between local ways of coming to know, interact with and use the land, and the dominant understandings and practices of global capitalism. Thus, IK can be used as a synonym for "traditional knowledge."

Unlike modernization theory, an IK position recognizes that traditions are anything but static. Rather they continually change and evolve over time, as cultural groups innovate, borrow and adapt their traditions to changing circumstances (Berkes and Folke 1998:4-5), be they ecological or social. The fact that indigenous traditions may have changed since contact with capitalist culture does not, therefore, imply that a study of those traditions is now irrelevant. Rather, an IK approach is consistent with the emphasis proposed above concerning the importance of focusing upon local level organization, and the type of concrete local knowledge which makes adaptive behaviour possible.

Yet there is more than one approach to the study of indigenous or local knowledge. In fact, much of the literature on participatory research and indigenous knowledge follows neither the cultural scale critique, nor the critiques of economic hegemony and modern science.

For this reason, the traditional ecological knowledge (TEK) perspective (which is best viewed as a more specific focus within the larger IK literature), has suggested the need for a more explicitly ecological approach to the study of traditional knowledge. Berkes defines TEK as "a cumulative body of know-ledge, practice, and belief, evolving by adaptive processes and handed down through generations by cultural transmission, about the relationships of living beings (including humans) with one another and with their environment" (1999:8)[143]. Like eco-holism more generally, TEK focuses upon both social and ecological patterns of relationship, as well as upon the relationship between the two. The twin focus upon both local social ecology, and local epistemology makes a TEK approach particularly suited to the search for the "appropriate epistemology of survival" proposed above. Such an explicitly holistic approach also allows TEK to avoid many of the criticisms which have been made of more conservative approaches to IK research, and provides support for more "radical" (or, as I prefer, more realistic) approaches, such as that proposed by Brouwer (1998:351).

The approach to IK research proposed by Sillitoe, for example, does not appear to support local economic or political autonomy in

---

[143] Berkes' definition of TEK was adopted by the Canadian government's Royal Commission on Aboriginal Peoples as its official definition of indigenous knowledge.

any form. Rather, it studies local understandings in order to "link them to scientific technology," as well as to contribute to "positive change, promoting culturally appropriate and environmentally sustainable adaptations acceptable to the people as increasingly they *exploit their resources commercially*" (1998:224, emphasis added). While he adopts the politically correct terminology concerning "participation," the "environment" and "sustainability," therefore, he fails to highlight the maladaptiveness of commercial scale culture itself. Instead, he accepts assimilation through commercialization and commodification as given, picturing "development" as an economic incorporation of local peoples into the global economic system.

In fact, Sillitoe's position would appear to suggest that anthropology's proper role is that of an *agent* of assimilation. For as he proposes, "the idea of harnessing anthropology to technical knowledge *to facilitate development* puts the discipline where it should be, at the centre of the development process" (1998:226, emphasis added). The suggested role for anthropologists in IK research is to act as "knowledge brokers" (1998:247), who are trained to mediate between cultures. The type of "development-oriented indigenous-knowledge work" which Sillitoe advocates, however, makes no pretense of "understanding others as they understand themselves," or of understanding indigenous epistemology. Rather, its goal is to understand and interpret "other cultures and their environments *as the demands of development require*" (1998:229, emphasis added).

Just as he fails to question the economic and political status quo, Sillitoe also fails to question the dominant scientific world view of Western culture. Indeed, Sillitoe dismisses the arguments of "anti-positivistic social scientists," who seek to "undermine" natural scientists without further comment (1998:232). Instead, he suggests that the real opportunity which an alliance between ethnographers and development scientists presents is "to compare indigenous statements and explanations against scientifically measurable data" (1998:227). Sillitoe does admit that "the heretical idea is gaining currency that others may have something to teach us" (1998:227). This is a point often made by TEK researchers, as when Colin Scott suggests that "local experts are often better informed than their scientific peers about local ecological conditions" (1996:71). For Sillitoe, however, it seems clear that traditional Western science remains the final arbiter of the validity of knowledge (Brouwer 1998:351). As he states:

> The implication of considering indigenous and scientific perspectives side by side is *not*...that scientists need to revise their working suppositions regarding objectivity, positivism, reductionism, and so on, to accommodate other views (Sillitoe 1998:226, emphasis added).

Such a statement implies that the type of reductive knowledge consistent with centralized planning is the final arbiter of the validity of local knowledge, or that abstract knowledge stands in judgment over concrete knowledge[144]. Indeed, a debate over the inclusion of TEK in an environmental impact assessment in Canada's North West Territories, which occurred in the mid 1990s, illustrates this contrast rather well. The debate was a response to the Government of Canada's Environmental Assessment Panel, which suggested that TEK must be given equal consideration with scientific research in assessing the environmental and socio-economic impacts of a proposed diamond mine in the area

---

[144] For a more detailed discussion of the contrast between IK and TEK, and their epistemological, political and policy implications see Dudgeon and Berkes (2003). Berkes (1999:17-58) provides a good summary of the approach's intellectual roots, as well as discussing various other issues of relevance.

(Howard and Widdowson 1996:34)[145].

On one side of the debate are its instigators (Howard and Widdowson 1996, 1997). They defend the status quo by arguing that TEK is unscientific, and should not be made a mandatory part of environmental impact assessments. Their argument is twofold. First, following the usual dualistic metaphysic, they suggest that TEK is not scientific, but spiritual, and that mandatory consideration of traditional knowledge is an "imposition of religion upon Canadian citizens" (Howard and Widdowson 1996:34). Further, they suggest that "spiritualism is obviously inconsistent with scientific methodology." They conclude, therefore, that TEK actually "hinders rather than enhances the ability of governments to more fully understand ecological processes," since there is no way in which "spiritually based knowledge claims can be challenged or verified" (Howard and Widdowson 1996:34).

By equating local knowledge with "spiritualism," then, and by invoking the duality between science and religion, they attempt to dismiss the relevance of concrete knowledge entirely. This leads to their second line of argument, which suggests that allowing Aboriginal peoples to control their own resource management strategies leads to a "conflict of interest." As they state, "Aboriginal peoples are the main harvesters of renewable resources in the north; clearly they should not be placed in a position of regulating

---

[145] This debate is also an excellent example of the growing influence of TEK on government policy, at least in the Canadian context. The importance of TEK was also promoted by the Canadian government's Royal Commission on Aboriginal Peoples (RCAP 1997), an arm's length commission which had the mandate of studying the social, political and ecological issues affecting the Aboriginal peoples of Canada, both through commissioning academic studies, and through an extensive series of public consultations with indigenous peoples and organizations themselves. In fact, the RCAP provided the first formal introduction of the Canadian policy community to both the concepts of "co-management and traditional knowledge" (Berkes and Henley 1997:29).

these resources" (1996:35). Thus, they argue for the superiority of centralized planning and the type of abstract knowledge supplied by Western scientific specialists over local knowledge and management.

On the other side of the debate are arguments advanced by academic proponents of TEK, which are more consistent with an eco-holist point of view (Berkes and Henley 1997; Fenge 1997; Stevenson 1997). They argue not only for the importance and value of TEK, but also suggest that the scientific community–and resource management planners in particular–may have something to learn from a study of the local knowledge contained in indigenous peoples' resource management systems. Indeed, Stevenson suggests that compared to the type of "impression management from a distance" of central planners, "Aboriginal peoples are in the *best* position to determine what hunting activities are sustainable or not" (1997:27, emphasis added), thus defending the importance of direct knowledge. Like Berkes and Henley (1997), Stevenson argues for the importance of local participation in resource management planning, which is consistent with adaptive management's emphasis upon the importance of "learning-by-doing" (Holling et al. 1998:58-9).

Further, Berkes and Henley argue in favour of "the recognition of indigenous knowledge as a legitimate source of information and values" and, in response to Howard and Widdowson's first claim, accuse them of subscribing to "the simplistic view that there is such a thing as objective, value-free science" (1997:30). Similarly, Stevenson claims that their article suggests that modern Western science supplies a "universal truth or methodology of observation, interpretation, wisdom or understanding" (1997:27). Both arguments appear more consistent with an eco-holist perspective, and with the goal of radical pluralism.

Though Howard and Widdowson explicitly deny that they are promoting the idea of science as supplying a "universal truth," or as being "totally objective," their reply to the above criticisms only serves to reinforce the fact that they are (1997:46-7). For as they suggest, "there are not different ways of knowing. There are different beliefs about the same phenomena" (1997:46). Clearly, the only legitimate way of "knowing" which they recognize is the reductive methodology of traditional science, while local knowledge is equated with "belief" or "spiritualism"[146]

In the same way, Sillitoe also appears to argue that science is the final arbiter of the validity of indigenous knowledge. After all, while Sillitoe clearly recognizes the value of indigenous knowledge, at least insofar as it contributes to the "development process," his understanding of science, and of its role in development, differs little from that of Howard and Widdowson. As Sillitoe suggests, for example, "the perspective of natural science has proved *successful* in promoting the kinds of interventions that *development demands*" (1998:226, emphasis added). Yet whether the type of interventions which development demands have proven to be successful in the *ecological* sense discussed above–the question of whether they are ecologically *appropriate*–is a question which he never raises. Sillitoe's reasons for not questioning the scientific orthodoxy concerning development issues, however, are quite clear. As he states it himself, "it is unrealistic to think that the scientific com-munity

---

[146] Widdowson and Howard (2006) maintains the same. Interestingly, even while citing Dudgeon and Berkes (2003) and several similar articles, they still claim that "this literature...has not really responded to our criticisms." In fact, however, the reverse is closer to the truth. Widdowson and Howard simply refuse to acknowledge that defenders of TEK advance a new understanding of science which soundly criticizes the antiquated position they continue to defend, which equates science as such with both *modern* science and its methods, as well as with universal truth.

could be persuaded to abandon its successful orthodoxy; it would be unable to make sense of the world without it" (1998:232).

Such a position assumes two things. First, that scientific orthodoxy is "successful" in the purely economic sense discussed above. Second, that there is no alternative to dominant understandings of science in general, and development science in particular. Ecoholism's critique of the scientific orthodoxy, however, challenges both of these assumptions, as does recent TEK research. Once again, this view calls for a paradigm shift away from mechanistic understandings of science, and towards more organic models (Berman 1984), which can adequately encompass an understanding of the unique characteristics of complex organic systems.

The most pragmatic reason for this shift is that "[t]here is a worldwide crisis in resource management because the existing science that deals with the issue seems unable to prescribe sustainable outcomes" (Holling et al 1998:352). Obviously, when applied to resource management, mechanistic and organic views have very different implications for scientific practice. The reductionist model is based upon utilitarian premises, which view nature as a collection of commodities which have no value until humans make use of them. The conservation practices suggested by this tradition have, since the 1930s, tended to rely upon a calculation of the "maximum sustainable yield" for any particular resource (Holling et al. 1998:345).

The primary focus of this approach is to successfully predict and control the abundance of those commercially valuable resources which are harvested. It is also reductive in two different senses. First, this approach tends to concentrate upon individual, commercially valuable species, in isolation from the larger ecosystem within which the population is embedded. Second, it is reductive in the usual way, in that it reduces the question of sustainable management to an equation which seeks to calculate the maximum possible harvest. Thus, in adopting both the ethics of maxima, and a reductive methodology, this approach to resource management science appears to have adapted itself rather more effectively to the monetary calculus of economics, and its methods, than to the nature of the ecosystems which it exploits, as the collapse of the cod fishery on Canada's eastern coast (among many other examples) appears to illustrate (Rogers 1995).

The synthetic and relational approach to resource management proposed by proponents of TEK and other related literature, on the other hand, is consistent with the development of a type of "adaptive management" (Holling 1978, Holling et al. 1998, Gunderson et al. 1995). This approach seeks to adapt not only resource management, but human institutions, techniques and values to the larger system of which they are a part. It also puts a greater emphasis upon "ecosystem processes" than upon "ecosystem products" (Berkes 1999:76). Consequently, adaptive management is *adaptive* in the sense that "it acknowledges that environmental conditions will always change, thus requiring management institutions to respond to feedbacks by adjusting and evolving" (Berkes 1999:60). One of the premises of this view "is that knowledge of the system we deal with is always incomplete. Surprise is inevitable" (Holling et al. 1998:346). As Berkes points out, "like many traditional knowledge systems, adaptive management assumes that nature cannot be controlled, since uncertainty and unpredictability are characteristics of all ecosystems" (1999:60).

There are two main reasons for this unpredictability in complex organic systems, both of which mitigate against prediction and control in an ecological context. First, the complexity of socio-ecological systems is such that we can never have *complete* knowledge of all of the relevant variables. Even in terms of eco-systems alone, science has never completely documented all extant species (especially in very diverse ecosystems, such as tropical rain forests and corral reefs), so that even rates of extinction generally have to be given as estimates (Quammen 2001). Similarly, many organisms, such as bacteria and viruses, evolve and adapt at an incredibly rapid rate due to their rapid rate of reproduction (Robbins 2002: 224-300). And one can hardly expect to be able to predict a system when one is not even aware of all of its constituent components, and their characteristics, for this implies that there are many potentially important relationships of which one is not even aware.

Second, an important characteristic of complex organic systems is their "non-linearity," in the sense that the same stimulus, applied to the same system at a different time, may produce strikingly different results. If one says "Hello" to the same person on a different day, and under different circumstances, for example, the response is far from predictable[147]. This is because the *context* is not the same. The same is true of actions affecting other complex organic systems, such as local ecosystems, or even the biosphere itself. Chris Bright, suggests that complex organic systems are unpredictable in three principle ways. First, they are subject to "discontinuities," or to "abrupt shift[s] in a trend or a previously

---

[147] Besides their non-linearity and resulting unpredictability, complex systems are also characterized by organization at a variety of different scales simultaneously, with complex feedbacks between them, and by self-organization (Berkes et al. 2003:5-7).

stable state"[148]. Second, complex organic systems are also subject to "synergistic" effects, wherein several seemingly unrelated phenomena combine in unexpected ways to produce an effect much greater than would have been expected from adding up their effects taken separately[149]. Finally, "unnoticed trends," such as ozone depletion from synthetic chemicals, or the introduction of invasive species, may also do a lot of damage before they are even detected by science (2000:9)

As a result of the inherent unpredictability of complex organic systems, what we need is *a science of complexity and surprise*, and this is why "adaptive management...treats policies as hypotheses, and management as experiments from which managers can learn" (Holling et al. 1998:358). Or as Berkes puts it, in adaptive management, "uncertainty and surprises become an integral part of an anticipated set of adaptive responses" (1999: 178).

It is for this reason that proponents of adaptive management are also commonly proponents of TEK, or of the possibility of learning more adaptive resource management practices from the values, epistemologies and practices of non-Western cultures. For not only have both economics and mechanistic science proven themselves to be ineffective tools for designing sustainable

[148] According to recent paleoclimatology, for example, global climate appears to be "bimodal," with relatively rapid flips (on a geological time scale) between ice ages and relatively warm periods, such as we are experiencing at present (Calvin 2000). The same pattern of relatively rapid transformation following long periods of relative stability is also proposed in evolutionary biology by punctuated equilibrium theory (Gould and Eldridge 1977), and appears to characterize long-term cycles of ecosystemic change as well (Gunderson et al. 1995).

[149] Bright (2001) provides the example of the collapse of the Black Sea ecosystem, which resulted from a combination of a drop in silicate content from the building of the Iron Gates dam (which lead to a collapse of the diatom population, the basis of the foodchain), the introduction of an invasive species of jelly-fish, and nutrient pollution from sewage and artificial fertilizers (and the algal blooms which resulted).

management practices, many indigenous peoples seem to have developed socio-ecological systems which are (or were) much more sustainable than our own. Successful indigenous resource management systems, when viewed as a "knowledge, practice, belief complex" (Berkes 1999:13-4), may thus be viewed as experiments in successful living, and drawing upon a knowledge of these alternatives may "speed up the process of adaptive management" (Holling et al. 1998:359).

As a variety of recent students of TEK have suggested (Berkes 1999; Johannes 1989; Williams and Baines 1993), there are also many similarities between traditional or indigenous management systems and the type of adaptive management which is proposed. So much so that some suggest that adaptive management itself is best viewed as "a sort of rediscovery of principles applied in traditional socio-ecological systems"[150] (Holling et al 1998:358). As Purcell points out, there is an increasing amount of literature which recognizes "the potential contribution of indigenous knowledge to the creation of a sustainable world" (1998).

In its attempt to learn about successful resource management from non-Western peoples, TEK again betrays the legacy which it owes to the earlier eco-holist literature, which has long demonstrated a willingness to learn about ecological patterns of living and thinking from non-Western cultures. In this vein, for example, Capra (1975) studied in detail the similarities between the emerging non-mechanistic scientific paradigm and taoism. Similarly, deep ecologists have often drawn explicit parallels between their own position and Buddhism (Devall and Sessions 1985:100-1; Fox

---

[150] Other recent studies have examined ecosystem-like concepts in indigenous cultures (Berkes et al. 1998), the importance of TEK to biodiversity conservation (Berkes et al. 1995; Gadgil et al. 1993), and its possible contribution to discussions of sustainability (Preston et al. 1995; Williams and Baines 1993).

1990:11-12; Zimmerman 1994:313-17). As noted in the introduction, many other ecological thinkers have also begun to explore the possibility of learning something about ecological understandings from an examination of the traditional views of Native Americans and other Aboriginal Peoples. Perhaps the most comprehensive survey to date is provided by Callicott (1994), with his consideration of "ecological ethics" from around the world.

What TEK research adds to this tradition–which has largely remained focused upon ideological, ethical, epistemological and political issues–is its clear emphasis upon practical matters such as resource management and biodiversity conservation. Besides the centrality of direct knowledge and appropriate scale in indigenous systems, perhaps one of the more important lessons which indigenous approaches to the management of resources provides is the fact that indigenous management systems tend to be *non-dualistic*. In other words, not only are human societies considered to be a part of nature, *values* are also explicitly incorporated into the system.

This is true of Cree resource management, for example, which was discussed briefly in Section II:3, and which is a useful case study in the present context as well. Not only is the Cree world view characterized by a belief in the spirits of animals and other natural phenomena but, as mentioned, by a belief in animal "bosses" which control the abundance and distribution of game, and which direct individual animals to be taken by hunters (Brightman 1993; Speck 1977; Tanner 1979). Similarly, as Cajete suggests of Native American peoples more generally, "[t]here was a widespread belief that each animal had a spirit village to which they returned and reported their treatment by humans, which henceforth directly affected whether that animal species would in the future give its life for humans" (2000:161). Consequently, success in hunting is closely associated with maintaining a proper attitude of respect towards the animals taken in the hunt, in order to assure success in the future.

Among the Cree this attitude of respect towards the animals hunted was expressed in a wide variety of ways; from the respectful attitude of hunters themselves, to the manner in which the animal is approached, killed and carried back to camp, to the manner in which it is butchered, consumed and its remains disposed of (Berkes 1999:83-7; Niezen 1998:26-8)[151]. As Berkes suggests, "[t]he rule about an attitude of humility is both important and universal. Hunters should not boast about their abilities. Otherwise they risk catching nothing" (1999:84), for animals will not offer themselves if they are not respected. Thus, it is not the hunter who is thought to control the success of the hunt, rather:

> A hunter always speaks as if the animals are in control of the hunt. The success of the hunt depends upon the animals: The hunter is successful if the animal decides to make himself available. The hunters have no power over the game; animals have the last say as to whether they will be caught (Cree Trappers Association 1989:21, as cited in Niezen 1998:27).

Among the Cree, animals are often described as "giving themselves" to the hunter, and maintaining an ethic and practice of respect for the animals is part of maintaining a proper pattern of reciprocity between the human and animal worlds. As Berkes suggests, "Cree social values such as reciprocity apply to human-animal as well as to social relationships" (1999:79). Where animals give themselves to hunters as a source of food, hunters must pay their "debt" for this sacrifice through properly respectful attitudes and behaviors. Thus, Cree beliefs concerning proper human-animal relationships not only recognize the inherent unpredict-

---

[151] For a discussion of the ethics of goose hunting among the Moose Cree see Scott (1996: 81-4).

ability of organic systems, but include humans within an essentially ethical pattern of reciprocal exchange with the species exploited (Preston 2002:194-218).

Once again, therefore, traditional ecological knowledge must be viewed as a "knowledge, practice, belief complex," in which spirituality, practical empirical knowledge and hunting practice all interact. Rather than dismissing the importance of spirituality, then, TEK researchers distinguish between the type of "figurative" knowledge which spiritual traditions provide, and the type of "literal" knowledge provided by direct experience (Scott 1996:72). They consider not only the implications of both, but their influences upon one another, and upon hunting practice as well[152].

Cree resource management practices have also been compared to adaptive management and found to compare favourably (Berkes 1999:126). As mentioned above, traditionally the Cree divided their winter hunting areas into a family territory system, each of which was supervised by a respected hunting leader or "tallyman" who had a long history of residence there, and who served as a "steward" for his territory (Niezen 1998:16). As Berkes describes the role of such hunting leaders, "[t]he senior hunter is the observer of nature, the interpreter of observations, the decision-maker in resource management, and the enforcer of roles of proper hunting conduct" (1999:89). Thus, Cree tallymen were responsible for determining who could join their group, for monitoring the location and abundance of game, and for keeping a mental inventory of the harvesting activities which took place on a

---

[152] Scott proposes this distinction in order to "get beyond the artificial dichotomy that separates Western and non-Western forms of knowledge," where Western knowledge is considered "scientific," and therefore "legitimate," and non-Western knowledge is dimissed as "mythic" (1996:70), as it was by Howard and Widdowson above. Cajete's (2000:28-31) contrast between the "rational" and the "metaphoric" mind draws a similar contrast.

year to year basis. In practice, this system promoted conservation in a number of ways.

First, it promoted *continuity in knowledge* both of the particular territory, and of the hunting activities which took place there. This was true not only on a year-to-year basis, since the same person continued to lead the hunt, but also intergenerationally. For as Niezen (1998) suggests, tallymen traditionally selected a successor who was not only a skilled hunter, but who also had a long history of residing with them on their territory. This type of long-term, direct knowledge was the basis of the Cree system of sustainable exploitation, and allowed for management decisions to be made on the basis of continuous, direct feedbacks.

Second, the family territory system also limited "the number of hunters who [could] operate in the family territories" (Berkes 1999:49), and allowed the numbers of people hunting in any particular territory to be adjusted to the availability of resources on a seasonal basis[153]. Third, in any given year, one or more territories may go unused, as when a tallyman thinks his territory has been over harvested, or is in need of a "rest." In that case he may ask to join another group while allowing a season for resources to recover in his own territory (Tanner 1979).

Finally, Cree resource management within the family territories, and more generally, also practices an ongoing *rotation* of the areas exploited in any particular season. As with the "resting" of an entire territory, this is akin to fallowing in agriculture. Scott (1996:78), for example, points out that while not tied to the territory system, the annual goose hunt among the Cree in

[153] Indeed, as Berkes and Fast (1996) suggest, while contemporary Cree populations are increasing rapidly, the contemporary family territory system serves as a check on the absolute numbers of hunters who actively harvest the land, and serves to stabilize the total hunting pressure produced by the community by regulating access.

northern Ontario is also lead by experienced hunters, who rotate the areas harvested out of necessity. This is because geese are highly mobile, and are not as likely to return to an area where they are hunted regularly. Thus, the rotation of the areas exploited is both adaptive and necessary for hunting success, so that the activities of hunters do not become predictable to the geese.

Berkes (1999:112) documents the importance of such rotation both in hunting and trapping in family territories, and in Cree fishing practices. Among the Cree, fishing is mainly a summer activity, and any particular area or lake would only be intensively fished on a rotating basis, and only returned to after a number of years had elapsed. As Berkes suggests, rotation, combined with the use of a variety of different mesh sizes in the nets used, results in a "thinning of the population" while conserving the "resilience" of the population exploited. Periodic thinning of the fish populations also "increases productivity by stimulating growth rates...in the remaining fish and helps the population renew itself" (1999:125). Thus, rotation and a mix of mesh sizes not only help to conserve the species exploited over the long-term, but also help to maintain both the productivity of the resource, and the stability of the population exploited.

The same is true of the rotation of the areas exploited on a seasonal basis on family territories in the winter months, when beaver is the main focus of the hunting effort. Rotation of areas harvested for beaver generally follow a four year rotation, being intensively harvested in one season, and "rested" for about three years. This helps to maintain the beaver population within an optimum range, as well as maintaining productivity. The "resting" of the area is necessary to allow beaver populations to recover, but periodic harvesting prevents them from reaching population densities which would deplete their food supplies and result in a temporary collapse in population numbers. Thus, as Berkes documents, an area should not be rested for too long, "for not only overuse can lead to a drop in productivity, but in the Cree world view, so does underuse" (1999:88)[154]. According to Cree understandings, "continued proper use of resources is essential for sustainability" (Berkes 1999:119)[155].

Besides illustrating the importance of direct, long-term knowledge as the basis for sustainable patterns of resource use in practice, Cree resource management also exhibits many similarities with adaptive management. Indeed, Scott suggests that the "Cree paradigm," or understanding of nature, pictures it as "a sentient, communicative world" (1996:76), and compares it to Bateson's (1979) discussion of the importance of information in complex organic systems, or more specifically, to his discussion of "the pattern which connects" (Scott 1996:72).

Geese, for example, are characterized by experienced Cree hunters as both adept at learning, and as able to communicate with one another. Flocks are observed to follow one another's flight paths over areas which they know to be safe, or free of hunters, and to safe landing areas. Consequently, the Moose Cree have worked out a sort of compromise in which hunting on the mainland is lead by experienced hunters similar to tallymen, who rotate the areas exploited so as not to become predictable to the geese, while setting aside certain islands for the unregulated use of part-time hunters in the community[156]. Flocks are also thought to

---

[154] Besides monitoring fat content, trappers also monitor "the health of the beaver-vegetation system...and other evidence of overcrowding, such as fighting among the beaver" (Berkes 1998:110).

[155] For Berkes' main discussion of the "importance of continued use for sustainability" see (1999:87-90).

[156] Which, as a result, have become areas geese are less likely to frequent.

be lead by older, more experienced geese, who have knowledge of previous hunting activity, much like humans (Scott 1996:77-81). Similarly, as Berkes suggests, the Chisasibi Cree are aware of the Inuit belief that caribou herds have "leaders," and that, consequently, you should "[n]ever take the first three caribou in the lead," but rather kill those who follow, so that the leaders will bring the herd back to the area in future (1999:106)[157]. Thus, in this view, other species are not only alive, but intelligent, responsive, and able to learn and communicate with one another. They are also able to adapt their behavior to the behavior of humans–which is very important knowledge if your livelihood is based upon successful hunting.

Cree resource management, not unlike adaptive management, also makes use of a more qualitative approach, rather than a quantitative one, such as traditional Western resource management tends towards. As Berkes suggests, the Cree do not even attempt to monitor absolute numbers of the species which they exploit. Rather, the central qualitative indicator monitored by the Cree, and by many other indigenous peoples, is the fat content of the animals killed. In fact, Berkes describes fat content as "a hunter's 'quality control' of the game" (1999:85). Monitoring fat content provides an indication not only of the health of individual animals, but is also a more general ecological indicator, which provides information on the availability of the resources which each species requires, and of the health of their habitat as a whole. Thus, unlike Western science, which "often gives priority to quant-

---

[157] This is a trait likely not uncommon among herd animals. When I was a child on my family's farm, for example, my father always kept one old cow rather than send her to market because she lead the herd to the summer pasture each spring and back again each fall–simply because she knew the way, and knew the seasonal routine, thus making the herd easier to direct and keep together during the moves each spring and fall.

itative population models for management decision-making," Cree resource management "uses a qualitative...model, which provides hunters with an indication of the *population trend over time."* In other words, it reveals "the direction (increasing/decreasing) in which the population is headed" (1999:109, original emphasis).

Monitoring fat content, then, is linked to the fact that the Cree, like Western scientific ecology, recognize that the populations of many species fluctuate over shorter or longer periods around an optimum population (Berkes 1999:82-3). Consequently, monitoring fat content provides for the possibility of "feedback learning" based upon direct knowledge concerning the direction in which a population is moving in these cycles, and allows hunters to adjust their harvesting practices accordingly. The ideal seems to be to harvest populations just before they reach their maximum in order to prevent a population collapse, and to return them to, or below, their optimum range, thus *stabilizing* population fluctuations over the long-term, and *optimizing* the productivity of the resource.

As mentioned in Section II:3, merely having a conservation ethic (and associated practices) does not guarantee that all members of a community, or even entire communities for that matter, will consistently follow it in practice. In terms of the internal diversity within communities, for example, it is the older and more experienced hunters who are most likely to be the most knowledgeable about practices consistent with such an ethic, while part-time or occasional hunters may follow more Western patterns, especially in the present day.

During fieldwork with the Missanabie Cree of northern Ontario, for example, I observed this difference while spending a couple of days fishing on Dog Lake, near the town of Missanabie, with members of the Missanabie Cree First Nation. While "traditional" Cree fishing made use of nets (and still does in more remote areas), fishing on Dog Lake was regulated by the Ontario Ministry of Natural Resources, so that we could only use the type of rods and reels typically used by Western "sports" fishers[158]. The men in my boat, who were also involved in the TEK study which I had come to assist them with, were practicing "catch and release" fishing, after the usual Western model of sports fisherman which I had learned myself while fishing with my father and uncles as a young man. In other words, our boat was throwing back the smaller fish in order to "preserve the stock," and keeping only the larger fish (or at least that was the intention, since we kept none of the few we hooked). My companions were also well educated in a Western sense, like myself, having three university degrees between them. They had also spent less time upon the land, actively harvesting it.

One of their cousins in another boat, however, who was tenting by the docks with his son in order to harvest fish for the use of his family, was using a different strategy. He had completed less formal education, to be sure, but had a reputation as a regular

---

[158] My primary purpose while visiting this and other communities in the summer of 1998 was to act as a consultant on each community's TEK studies; however, I also took the opportunity during each visit to engage in participant observation. I kept detailed field notes, not only concerning the progress of, and my participation in each community's research project, but also of my other observations and experiences in each community. The present incident is from my first visit with the leaders of the Missanabie Cree's TEK project, in which I spent several days touring sites on their traditional territory mentioned in their report—such as petroglyphs, archaeological sites, a grave site, a centuries old portage and the remains of a fur trader's cabin. We also spent time fishing, eagle-feather hunting and discussing the project with community members. The report itself was prepared three weeks later during a subsequent visit to their band office in Sault Ste. Marie, Ontario. Interestingly, while the fishing incident I describe was included in my field notes, I only arrived at my present interpretation of its significance while reading articles on Cree resource management some months after returning from the field.

harvester, who spent much more time on the land hunting and fishing. Rather than throwing back the small fish, he was keeping whatever he happened to hook, no matter the size. In fact, when we all returned to the dock after a day's fishing, my companions (who had returned empty handed, like myself) were even "ribbing" him in a joking manner about his fishing practices, and asking him why he was keeping such small fish. He merely smiled, laughed, shook his head, and continued filleting his abundant fish, but did not offer an explanation.

It was only later that I realized that his fishing practices were consistent with traditional Cree understandings of the proper ethics of respect for animals. I have often wondered since whether he was thinking–as he smiled and laughed–"if you guys hadn't spent so much time in school you'd know this." For as was discussed above, in the Cree world view, animals are thought to "give themselves" to the hunter. Thus, a fish which takes your hook has voluntarily offered you its life in order to sustain you and your family. To refuse the offer is thus a serious breech of the ethic of reciprocity and respect for the animals, since catch-and-release "sport" fishing amounts to merely "playing with fish" (Berkes 1999:118). In other words, it is equated with treating them disrespectfully. Consequently, those who practice it can expect less fish to offer themselves in future, and less success at fishing generally[159].

Aside from the issue of diversity within the community, Berkes (1999) also provides two examples of the manner in which a conservation ethic was suspended by an entire community, only to

---

[159] In fact, based upon my own observation, the "traditional" fisher and his son brought in far more fish than the three of us in my boat ever hooked. They were also more successful even when our boats were within shouting distance of one another, and were fishing the same area. Perhaps his strategy was working, though I have only two days of data on which to base that observation.

be reestablished after a period of time, each of which is instructive for a different reason. The first, mentioned above, is the collapse of the ethic of respect for the Caribou among the Chisasibi Cree early in the twentieth century (Berkes 1999:95-109). As the elders told the story, before the caribou disappeared from the Chisasibi area in the early 1900s, repeating rifles had been introduced into the area for the first time. This new technology lead to a breakdown of the conservation ethic, and as Berkes puts it, "previously respectful hunters, dizzy with newfound power over the animals, lost all self-control and indiscriminately slaughtered the caribou." They killed so many, as the elders told it, "and wasted so much food that the river was polluted with rotten carcasses" (Berkes 1999:102).

The following year, the hunters waited and waited and no caribou came. Back in the early 1900s, the elders explained that this was because they had treated the caribou disrespectfully. In other words, they blamed themselves for the collapse of the caribou population, even though there were likely other ecological factors involved from a scientific perspective[160]. But the elders at that time also made a prediction: they stated that one day the caribou would return to the area, but if the hunters did not treat them with proper respect, they would not stay.

About 70 years later, after the caribou population had recovered, and begun to extend their range back into the Chisasibi area, large numbers of caribou were killed for the first time in generations in the winter of 1983-84. At the time, community leaders were concerned, "not because large numbers were killed," but because of the manner in which the hunt was conducted. Many hunters had been shooting indiscriminately, "letting wounded

---

[160] Berkes suggests that the population had collapsed, for reasons which are not fully understood, and retreated to a smaller range outside of Chisasibi territory (1999:97-99).

animals get away, killing more than they could carry, wasting meat, and not disposing of the wastes in a respectful manner" (Berkes 1999:101-2). While the elders were concerned at this lack of respect, they said nothing at the time. Instead, they waited.

The next winter came and the caribou were far less abundant, and less accessible. They stayed away from the roads, and many hunters came home empty handed. This got the younger hunters concerned, and now the time was right for the elders and hunting leaders to come forward and "draw some lessons from the apparent reluctance of the caribou to come back to their lands" (Berkes 1999:102). A town meeting was called where the story passed down from 70 years previously, which many were familiar with, was repeated by a couple of the most respected elder hunters. This made the younger hunters realize their mistake without directly criticizing them. Instead, the elders waited until the prediction contained within the story came true, thus validating their traditional knowledge in the eyes of the community, and then retaught the proper ethic of respect for the caribou to the younger generations.

Consequently, the following year the hunt was conducted very differently. Now it was regulated at the community level by the Cree Trappers Association, whose members ensured that "the hunt was conducted in a controlled and responsible manner, in accordance with traditional standards" of respect for the caribou (Berkes 1999:103). Since that time the caribou have reestablished themselves throughout Chisasibi lands, all the way to the James Bay coast.

This incident clearly illustrates the central importance of both direct, long-term knowledge, and of continuity in knowledge to adaptive management, for as Berkes relates, "[e]lders provide the corporate memory for the group" (1999:95). It also illustrates the manner in which a conservation ethic, even while temporarily forgotten, can be relearned even after decades lacking any direct experience in harvesting a particular resource, through the teachings of oral history, and the traditional ecological knowledge retained by the elder hunters.

Another documented incident in which a conservation ethic appears to have been temporarily suspended among the Chisasibi, is the collapse of beaver populations in the 1920s (Berkes 1998:113, 1999:96). This incident raises important issues regarding property rights regimes and community level management for sustainability. The beaver depletions followed shortly after the construction of railroads into the southern James Bay area, which first opened the area to non-Native trappers "at a time when fur prices were high." This lead to a situation in which the Cree were "unable to regulate access," since outsiders did not respect their family territory management system, and as a result, their conservation ethic seems to have been temporarily suspended. In other words, "rather than allowing outsiders to take the remaining resource," Cree hunters joined them in over harvesting beaver, leading to a collapse of the population by 1930 in a classic example of the "tragedy of the commons" (Berkes 1999:96)[161]. When the outsiders left due to the lack of beaver, however, the Cree were able to successfully reestablish their resource management practices, and beaver populations rebounded by the mid-1950s.

In Hardin's (1968) original formulation of his thesis, of course, he had argued that commons management was *inherently* unsustainable. Yet he had also proposed that western utilitarian, or economic logic was universal, as when he suggested that pollution

---

[161] Consequently, this also illustrates the fact that human communities must adapt not only to ecological conditions, but also to one another.

was "inevitable" as long as we "behave only as independent, rational, free-enterprizers" (1968:1245). In other words, as long as we think only in terms of our own short-term self interest, and follow the ethics of maximization which utilitarian or economic logic promotes, both the over-use of resources and pollution make perfect sense, since individuals (or corporations) enjoy the benefits, while society, or nature, or future generations pay the costs. In other words, costs are "externalized." Consequently, Hardin concluded that "[f]reedom in a commons brings ruin to all" (1968:1242), and recommended that the tragedy of the commons could only be averted by a combination of private property (1968:1245) and "mutual coercion, mutually agreed upon" (1968:1247).

In other words, what Hardin suggested was that the causes of problems such as pollution, and the unsustainable use of resources, were to be found in freedom and common property. Yet both of these proposals have been questioned by more recent cross-cultural research on resource management and property rights regimes. First, the more recent literature distinguishes between common property regimes proper, where access is regulated by a *community* of users (as among the Cree) and the type of "open access" regime which Hardin described (Berkes 1999:142; Burger and Gochfeld 2000:130). And in light of the abundant cross-cultural evidence that common property can be sustainably managed at the community level, this literature proposes that it is actually the "open access regime," where there is a *complete absence* of regulation, which leads to the "tragedy of the commons."

Consequently, common property is not the cause of the problem. Yet neither is freedom, particularly at the community level. Rather, as Burger and Gochfeld suggest in their review of the relevant literature, "[c]ommons management is more likely to succeed when the users depend on the resource...share a common understanding of it...when their is grounds for trust" among the users, and when they are able to "form an *autonomous controlling body*" to manage the resource at the community level (2000:132, emphasis added). As the Cree case concerning the collapse of the beaver population suggests, they must also be able to effectively *exclude outsiders*. For as Berkes suggests, "[s]ocieties do not establish conservation rules and ethics for the benefit of outsiders" (1999:96).

From a cross-cultural perspective, therefore, "the phenomena in which the individual off loads costs to society while pursuing private benefits" (Berkes 1999:142), which Hardin identified, is a result of neither freedom nor common property. Rather, it is a result of open access situations and, even more importantly, a result of the utilitarian or economic logic which Hardin assumed to be universal, and its ethic of maximization. As my discussion of the principle of sufficiency, which guided many non-Western economic systems suggests, however, this logic is far from universal. In fact, the tragedy in question is more appropriately described as "the tragedy of *capitalism* in the commons."

Thus, as Berkes at al. suggest, "the alternative world views of traditional peoples could provide insights for redirecting the behavior of the industrial world towards a more sustainable path" (1995:299), by teaching us many useful lessons about sustainable resource management at the local or bioregional level. And while Sillitoe proposes that "the philosophy underlying indigenous-knowledge research is unexceptionable" (1998:224), the same cannot be said of TEK. This is primarily due to its explicit emphasis upon the importance of *ecological* knowledge and issues. This allows TEK research to draw upon several decades of theoretical development within the broader eco-holist literature, including its critiques of mechanistic science, centralized planning and economic logic.

Yet as Berkes reminds us, "much of contemporary ecological research uses reductionist thinking. Thus, not all ecological science is sympathetic to traditional ecology" (Berkes 1999:182). Among those whose views are informed by eco-holist thought, however, and by the distinction between deep and shallow ecology, there is much interest in "bridging" the two traditions. Berkes, for example, suggests that adaptive management and other holistic traditions within Western science "are potentially suitable to provide frameworks for integrating Western and indigenous knowledge," and–consistent with radical pluralism–this suggests that each should be recognized as legitimate forms of knowledge (1999:178-9). Indeed, contemporary Native American scholars also often emphasize the fundamental importance of ecological issues, or call for the need for "bridging" the two traditions. As Sioui proposes:

> If there is to be a positive development of history and of sciences in general (social and otherwise), *ecology must become the common mother of all sciences;* and history, like the social sciences, must consider the ecological dimension as fundamental (1992:101, emphasis added).

Cajete (2000), on the other hand, describes TEK as a "Native science" in its own right, which focuses upon the study of "natural laws of interdependence," just as eco-holism does. Yet as was also implied by the introductory quotation from his work, Cajete also suggests that "Native people were the first ecologists" (2000:207). Consequently, not unlike TEK researchers and eco-holists more generally, Cajete proposes that Western science has much to learn from Native science. Indeed, Cajete argues that "Western science *needs* Native science to examine its prevailing world view and culture" (2000:285, original emphasis), as does the eco-holist literature. Not unlike TEK researchers, Cajete also calls for the importance of bringing the two traditions together so that they may learn from one another, for "[c]reation always engenders synthesis." In fact, Cajete suggests that we are already "in the early phases of a synthesis between Indigenous science and certain aspects of Western science" (2000:208).

A similar sentiment was also expressed by the Moose Cree of northern Ontario in a recent report on traditional land use among their people:

> As Aboriginal peoples, we have a long history of observing and experiencing environmental changes, and if we–as Aboriginal people–put our knowledge together, we can explain some things which scientists cannot explain as well. At the same time, however, scientists can explain some things which we can't explain. Accordingly, we firmly believe that traditional knowledge is just as important as scientific knowledge...and that the two may provide complementary means of understanding the same problems. Imagine the 'know-how' which could be assembled if the scientific and traditional approaches to ecological knowledge were combined[162]!

Perhaps the most important lesson which Native science teaches us, however, is provided by its non-dualistic character, and the fact that non-economic values are explicitly incorporated into its investigations. Thus, despite the importance of appropriate scale, we should not expect our activities to be ecological simply because they are decentralized, nor simply because the scale of our organization, or the size of our tools and machines have been reduced to a more human scale. Rather, what both TEK researchers and recent Native American scholars suggest, is the central importance of long-term, direct knowledge at the local or bio-regional level. Further, such local knowledge must be guided by a

---

[162] This quote is taken from an unpublished report entitled "Moose Cree First Nation Traditional Land Use Mapping Project," which reported on the initial findings of their government financed but community organized TEK research project, which I assisted in editing into its final form in the summer of 1998.

deeply ecological sensibility, or by the knowledge that all things in nature are intimately interconnected, and that all are deserving of our respect. For as Standing Bear once suggested:

> the man who sat on the ground in his tipi meditating upon life and its meaning, accepting the kinship of all creatures, and acknowledging unity with the universe of things was infusing into his being the true essence of civilization. And when native man left off this form of development, his humanization was retarded in growth (1978:250)[163].

To return for a moment to the discussion of the different approaches within the IK literature, however, not all IK researchers directly support indigenous political and economic autonomy. Sillitoe's non-critical stance concerning the relevance of IK, for example, extends not only to the dominant scientific paradigm, but to questions of policy and politics as well. In other words, it extends to issues which might involve questioning the *aims* of development. As he suggests, while IK research "intimately and unavoidably involves political issues," in his opinion, anthropologists working in the area should merely "inform politicians and others about issues as they perceive them and leave the responsibility for policy decisions to them" (1998:231). In Sillitoes opinion, then, as anthropologists, we are "not politicians, management consultants, or policy makers," and any pretense towards such ends would test "the limits of our disciplinary competence" (1998:246-7).

The non-critical stance advocated by Sillitoe with regards to development policy has been criticized by others, however, both directly in response to the article in question, and in more general critiques of the IK literature. Both Stirrat (1998:243) and Posey (1998:241) suggest that Sillitoe has ignored much of the "political context" in which IK research is taking place. This is not surprising given the views quoted in the above paragraph.

---

[163] This quote is also reminiscent of Bodley's (1999) claim that "sapienization" is the main goal of domestic scale cultures.

In an earlier article, Arun Agrawal also raises the point that the central objective of much IK research is politically inappropriate, for reasons which are consistent with the above discussions of appropriate scale and direct knowledge. As Agrawal proposes, the earlier IK literature has tended to advocate the *ex situ* preservation of IK in centralized, bureaucratically organized databases, where it will primarily be of use to agents of development. Indeed, she claims that this is "not just the preferred strategy...it is almost always their only strategy" for preserving IK (1995:430). In other words, Agrawal suggests that IK researchers advocated the *external* preservation and exploitation of IK by development agencies, without a sufficiently strong emphasis upon the concurrent preservation of the indigenous cultures which produce it, nor upon the *ownership* of that knowledge by indigenous peoples themselves.

Indeed, while emphasizing the central importance of concrete knowledge and feedback learning, Berkes suggests that external preservation of TEK is not only inappropriate, but also ineffectual, since "[t]he written page will never be an adequate format for the teaching of indigenous knowledge. It can only be taught properly on the land" (1999:28, original emphasis). Further, from an eco-holist perspective, such an approach does not sufficiently problematize current political relationships between the global capitalist system and the local systems of politics and production which it continues to colonize.

Consequently, Agrawal suggests that "it might be more helpful to frame the issue as one requiring modification in political relationships" (1995:431). This suggests an alternate position which is more consistent with eco-holism, a position which advocates the *in situ* preservation of IK or TEK through changes to "state policies and market forces," and Agrawal offers a detailed discussion of the arguments for and against such a course of action (1995:432). For if successful adaptation requires direct knowledge and local level feedbacks, then *in situ* preservation is the *only* way of successfully preserving TEK, since "much of indigenous knowledge makes no sense when abstracted from the culture of which it is a part" (Berkes 1999:23). Such an emphasis upon the preservation of local cultures and production systems is also similar to that taken by Stone, in her comment on Sillitoe's work, particularly when she states that:

> Goodwill and inter-disciplinary open-mindedness will not be enough to change this highly bureaucratized system and world of development planning, implementation, and evaluation unless accompanied by institutional and policy change (1998:243).

It appears, that there is more than one approach to IK research even among those who use the term, which is not adequately encompassed by Sillitoe's contrast between the market-liberal and the neo-populist approaches. Neither approach sufficiently characterizes an eco-holist or TEK approach, even though the latter does advocate populism, or "bottom-up" research. Further, neither approach is explicitly ecological. Neither is a contrast between IK and TEK adequate—for not only is TEK a subfield of IK, but many of the above critics of more conservative IK approaches do not identify themselves with TEK[164].

Fortunately, Purcell (1998) provides us with a more useful suggestion with his contrast between the "indigenous knowledge approach" and the "ecoliberal" approach. The "ecoliberal" approach "aims at helping to integrate people into the capitalist market on their own terms" (1998:267), not unlike the approach which Sillitoe

---

[164] Both Agrawal (1995) and Purcell (1998), however, do make much more explicit reference to ecological issues than does Sillitoe (1998).

advocates. An "indigenous knowledge" approach, on the other hand, is based upon an understanding of the fact that "capitalist transformation threaten[s] local communities and ecological systems and is therefore unsustainable" (1998:265). This not only suggests that IK is consistent with an eco-holist or TEK approach, but implicitly suggests that Sillitoe's arguments should not be considered consistent with an IK approach as well. For while both might be consistent with the promotion of more participatory approaches, only the latter explicitly questions the aims of development and the policies which guide it, which are primarily economic. Purcell's definition of an IK approach is far more consistent with the critique of commercial scale culture and its ideology discussed above, for as he continues, "[i]nterest in the study and application of IK—or indigenous *praxis* as a transformative process—logically implies indigenous people assuming relative autonomy" (1998:260, emphasis added).

The TEK literature, with its explicitly ecological emphasis, therefore, appears to support many of the criticisms which have recently been made concerning more conservative approaches to IK research. And from an eco-holist perspective, the choice of methodology and focus for indigenous knowledge researchers is as much ethical and political as it is practical. Shall we study indigenous knowledge "as the demands of development require," or shall we promote its importance for the benefit of the peoples who possess it?

Purcell suggests the importance of either fully collaborative work, or the promotion of IK research conducted by indigenous peoples themselves, for their own use and benefit, rather than for the benefit of agents of development (1998:268). As he states, IK researchers "must choose between being facilitators of local autonomy...or be agents of hegemonic 'progress'" (1998:267). From an ecologically informed perspective, therefore, "it is economic development based on the logic of unbridled growth that destroys indigenous territories" (1998:265), along with the cultures, local production systems and systems of direct knowledge of local communities. Consequently, the choice between these two options is clear: to support local autonomy and local participation in resource management issues–whether among aboriginal peoples, or people generally–so that local cultures may successfully adapt to the great diversity of local ecological conditions in which they reside.

To conclude with a sentiment which both eco-holists and indigenous peoples, in their varying ways, have often repeated:

> We cannot adequately express our feelings of horror and revulsion as we view the policies of industry and government in North America which threaten to destroy all life. Our forefathers predicted that the European Way of Life would bring a Spiritual imbalance to the world, that the Earth would become old as a result of that imbalance. Now it is before all the world to see–that the life-producing forces are being reversed, and that the life-potential is leaving this land...We bring to your hearts and minds that right-minded human beings seek to promote above all else the life of all things...It is our duty as human beings to preserve the life that is here for the benefit of the generations yet unborn (Knudtson and Suzuki 1992:194)[165].

---

[165] This quote from a declaration of the Six Nations Iroquois Confederacy was first made public on April 17, 1979.

CONCLUSION:
ON UNITY IN DIVERSITY

An Indian looks at nature and sees beauty–the woods, the marshes, the mountains, the grasses and berries, the moose and the field mouse, the soaring eagle and the flitting hummingbird, the gaudy flowers and the succulent bulbs. He sees the diversity of the various elements of the entire scene...He surveys the diversities of nature and finds them good. An Indian thinks this might be the way of people...To the Indian this is the natural way of things, the way things should be, as it is in nature. As the stream needs the woods, as the flowers need the breeze, as the deer needs the grasses, so do peoples have need of each other, and so can peoples find good in each other (Cardinal 1969:78-9).

To exist in a creation means that living is more than tolerance for other life forms–it is recognition that in differences there is the strength of creation and that this strength is a deliberate desire of the creator (Deloria 1994:89).

IN THE FALL OF 1992, myself and two fellow anthropology students met with Neil Hall, a Native drummer and singer, to learn something about the meaning of contemporary powwow gatherings in our area. We had contacted Hall at the suggestion of a professor of Native Studies, and he had invited us into his home in Winnipeg's North End. Once we were seated around the kitchen table, and had introduced ourselves to one another, he asked us what we would like to know. I suggested that perhaps it would be best if he began by telling us what he, himself, thought to be important about pow-wow gatherings.

He responded, not by talking about powwows, at least not initially, but with the following remark, which I have carried with me ever since: "Our beliefs are not a religion, but a way of life[166]." He then went on to discuss the beliefs of his people, humanity's relationship to nature, and the importance of looking to the manner in which Nature itself does things when trying to under-stand how humans themselves should live. Thus, he immediately

[166] Neil Hall, personal communication, November 1992.

reminded us of the fact that any one cultural event cannot be adequately understood without also understanding something about its place within the larger culture, and that culture's conceptions of their relationship to nature.

Throughout the preparation of the present work, I have attempted to keep this sentiment in mind, and have used it as a touchstone to guide my approach. For not only does it correspond well with central themes of the Native American literature discussed above, it also expresses one of the central understandings of eco-holism; more specifically, its *holism*. Throughout the present work I have attempted to express this holistic sentiment (certainly less succinctly, and perhaps less well than Hall) by suggesting that the ideological sphere, the social sphere and the ecological sphere cannot be considered in isolation. Rather, it is the relationships between them which are the focus of study. Nor was I surprised, as my research continued, to find that Rodman had once described Leopold's land ethic in the following way: "[i]t is an 'ethic' in the almost forgotten sense of 'a way of life'" (as cited in: Rolston 1988: 353).

In other words, as was implied by Hall, whether one refers to them as religion, philosophy or ideology, one's beliefs are intimately related to one's way of life, and that way of life to the ways of all living things. And in the analysis of a culture, or a way of life, it is the most prevalent ideology–the one which is most consistent with its way of life–which is of central concern. Clearly, the dominant values in Western culture consist of the constellation of values embodied in economics, modern science and liberal individualism, which support an exploitive attitude in our relationship to nature. In many Native American cultures past and present, on the other hand, the predominant values appear to be more collectivist, and to retain an abiding respect for Nature and its ways, which is consistent with a more restrained, respectful and sustainable relationship with the natural world. This also appears to be a view which an increasing minority of Westerners have now come to share to a large degree. Both the world view and ethos which inform eco-holism are broadly consistent with those which have been expounded by Native Americans since the time of first contact. The reason for this is no mystery, for as Standing Bear related:

> The white man does not understand the Indian for the reason that he does not understand America. He is too far removed from its formative processes. The roots of the tree of his life have not yet grasped the rock and soil...The man from Europe still hates the man who questioned his path across the continent. But in the Indian the spirit of the land is still vested; it will be until other men are able to divine and meet its rhythm. Men must be born and reborn to belong. Their bodies must be formed of the dust of their forefathers bones...Every problem that exists today in regard to the native population is due to the white man's cast of mind, which is unable, at least reluctant, to seek understanding and achieve adjustment in a new and a significant environment into which it has so recently come (1978:248-9).

Consistent with this sentiment, it has been the purpose of the present work to encourage more eco-holists to turn directly to Native American literature, philosophies and practices in order to gain insights into both an ecological sensibility and sustainable lifestyles. After all, in the North American context, it is exactly the type of accommodation and adaptation of Western culture to the nature of the lands it so recently colonized, of which Standing Bear speaks, to which eco-holism now aspires. To my mind, it would appear that eco-holism should seek not only to balance the ends of humanity with those of nature, but also to balance the short-term needs and desires of individuals with the long-term needs of human communities and societies. And in light of the fact that Native American philosophies were born of this soil, to which we now belatedly attempt to adapt ourselves, eco-holism–*at least in the North American context*–should be an attempt to balance Native American philosophies with those of the Western world. The same is true in many other parts of the world with respect to their own Aboriginal populations.

Not only are ecological philosophies broadly consistent with Native American philosophies, so also are their criticisms of the traditional ends and dominant values of Western culture, and of their consequences in practice. The more decentralized, egalitarian pattern of organization which eco-holism recommends–which would restore local cultural diversity by "scaling down" the level at which economic and political systems are organized–would also restore a pattern of life and organization which is broadly similar to that which Native Americans and other indigenous peoples had evolved centuries if not millennia ago.

As the Western world has begun to rediscover a concern for the natural world, therefore, it has also tended to rediscover many patterns of belief which have been promoted by indigenous peoples for as long as we have had any knowledge of them. The next step, however, is to *embody* these patterns in our own ways of life, a task for which the economic ethos is clearly an inappropriate guide, for as LaDuke suggests, "There is no way to quantify a way of life, only a way to live it" (1999:132). As Deloria reminds us, therefore, reiterating the themes raised by Hall and Standing Bear above, "[t]ribal religions are actually complexes of attitudes, beliefs, and practices fine-tuned to harmonize with the lands on which people live" (1994:70). Perhaps, then, it is time we began to listen, and to learn from those who have had the most experience with these ways of thinking and living–and with this land–particularly if adaptation is our goal. In other words, it is time we stopped treating the land as an "invader" and began to learn to live as "natives" on the land to which we so recently came (Dasmann 1988).

Ecological concerns, therefore, appear to provide a point where the cultures of Native America and those of the Western world can begin to come together, as the discussions of local knowledge and Native/ecological resistance illustrated above. In other words, ecological concerns provide a point of balance which can contribute towards transcending the traditional antagonisms which have separated the two cultural types through pursuing a common goal which is of equal relevance to all peoples–*survival*. Perhaps our common ecological problems may allow us to recognize a point of common ground at last, or as Brooke Medicine Eagle expressed this point:

> The Indian people are the people of the heart. When the white man came to this land, what he was to bring was the intellect, that analytic, intellectual way of being. And the Indian people were to develop the heart, the feelings. And those two were to come together to build the new age, in balance, not one or the other...I think we're beginning to see the force of this land, that receptive force, come back again, and that balance is beginning to happen (as cited in: Halifax 1979:90-1).

While other contemporary Native Americans may take exception to the contrast between the heart and the intellect as characteristic of the two cultural types, the emphasis upon the importance of balancing the strengths of the two is clear. After all, we cannot simply go back to hunting and gathering or horticulture. Social and ecological conditions have changed dramatically from what they were in the past, largely due to the expansion of the capitalist system, and we are now confronted by a very different context. Thus an ecological lifestyle of the future must take advantage of the strengths of each tradition, in order to create philosophies, ways of life and technologies which are mutually beneficial to all involved, including the Earth itself.

From the West comes the scientific tradition, while from Native America comes the long-term, local level knowledge of ecological systems which appears to be so essential to successful adaptation. The egalitarian attitudes consistent with respect for nature, and the principle of sufficiency, seem equally crucial to the establishment of sustainable societies. Indeed, as Naess observes with regard to Native American philosophies, "[t]heir conception of the human situation is more realistic than that offered in our techno-natural scientific education" (1989:175)[167].

So both cultural types have a contribution to make if we are to restore balance to the creation. From the West comes technological knowledge or "know-how," which has always been the strong point of applied science. From Native America comes the values, sensibilities and local ecological knowledge which must guide the development and application of appropriate technology. Thus, as Deloria expresses this point, in terms which some may find preferable to those used by Medicine Eagle, where "[t]he Western hemisphere produced *wisdom*, western Europe produced *knowledge*" (1970:11, emphasis added).

It is certainly the height of irony, of course, that it is to the very peoples whom it has spent so much time conquering, missionizing and attempting to assimilate that the Western world must now turn for the wisdom which it so desperately needs–the very wisdom which it so zealously attempted to eliminate. We should be thankful that the attempt failed, and that many First Nations have retained both their traditions, and their identities as separate peoples. After all, it is only through the coming together of their

---

[167] While he upholds a Native American outlook as "more realistic" than that of modern science, however, Naess undertakes no discussion of Native cultures, and references none of their literature.

wisdom–or something very much like it–with the technical knowledge of the West, that ecological problems are likely to be solved. As N. Scott Momaday, the Kiowa novelist once observed:

> the Indian has assumed a deep ethical regard for the earth and sky, a reverence for the natural world that is antipodal to that strange tenet of modern civilization that seemingly has it that man must destroy his environment. It is this ancient ethic of the Native American that must shape our efforts to preserve the earth and the life upon and within it (1976:13).

Eco-holists, therefore, are not the only ones who recognize the importance of bringing Native insights to bare on ecological problems, for as was discussed above, many contemporary Native Americans appear to agree. As McGaa states, for example, "I call on all experienced Native American traditionalists to consider coming forward and sharing their knowledge. Come forth and teach how Mother Earth can be revered, respected and protected" (1990:vii). Indeed, the sentiments expressed by eco-holists, on the one hand, and by contemporary Native scholars, on the other, appear to represent complementary aspects of the same thought. And though we may begin construction on opposite sides of the river, as it were, increasingly, both schools are attempting to construct the same bridge, and our efforts may meet in the middle. For as Cajete suggests, "[b]ridge building is underway, and both sides stand to gain from the dialogue" (2000:285). Thus, in the struggle to build an ecological future, the ecology movement and First Nations peoples are natural allies, who should learn from one another, and seek one another's assistance, and I hope that the present work may add a further stone to this bridge.

Yet besides serving as a bridge between these two cultural traditions, the bringing together of ecological concerns and Native American traditions is also important for another reason. For one of the great values of this type of cross-cultural approach is that it illustrates the fact that people have, beyond doubt, lived in the more egalitarian, decentralized, participatory manner which ecology now recommends. While some might invoke the values of the elite, and attempt to dismiss its recommendations as "uneconomical," therefore, they cannot be dismissed as unrealistic, non-feasible, nor as utopian. After all, these patterns of life are consistent with what may be the original pattern of human beliefs and lifestyles. The contemporary practices of the Western world, on the other hand, also come to be seen for what they are–a very recent experiment which has proven itself to be massively destructive, inequitable and inadequate to ecological realities. So as Cajete suggests:

> Human domination has not sustained the life of our Earth, and we must learn what will...The cosmology of human progress and domination over nature is no longer viable, sustainable, or ultimately desirable...We are the Earth consciousness at a new place realizing the consequences of choices made by only a small portion of the human race (2000:281).

Yet given that the earliest human societies were also domestic scale cultures, as were those of most of Native America, these types of philosophies, at least in their general outline, may also fairly be considered to be representative of the original views of the human race. Thus, rather than equating human nature with the peculiar and recently invented ideals of the capitalist elite, it would be fairer to equate human nature with the types of philosophies which characterized the vast majority of human societies throughout the vast majority of human history–those of domestic scale cultures. Such a position would also suggest that these were views which were once shared by Europeans as well, for several ecoholists have suggested that, while surely dualistic and hierarchical in their thinking, even pre-capitalist Europe retained much more organicist views of nature prior to the advent of mechanistic thought (Berman 1984; Merchant 1980). The philosophies and lifestyles of indigenous peoples thus represent an enduring critique of the Western world, simply because they prove beyond doubt that more ecological lifestyles *are* possible, as are the philosophies or world views which support them.

Nor are questions of appropriate scale, and of the choice between prioritizing economic versus ecological values merely abstract philosophical questions. Rather, these issues have practical consequences on the ground. Once again, this suggests the importance of radical pluralism in practice. For the present work has essentially been an exercise in *unity-in-diversity*, which has attempted to illustrate the internal consistency of the various strands of the eco-holist school, and of the various varieties of Native American philosophy, as well as the commonalities which crosscut the two streams of thought. Even though Native American literature and eco-holist thought express their philosophies in different terms, from the perspective of radical pluralism we must recognize that they are essentially speaking the same language, and share many of the same goals–such as local political autonomy and local economic self-sufficiency and sustainability–and the same enemies.

Indeed, even though the concepts of "unity in diversity" and "radical pluralism" are central relational symbols proposed by the eco-holist literature, the implications of this position are also beautifully expressed in the introductory quotations from the works of Cardinal and Deloria. Both give expression to a relational symbolism in which opposites are not seen as separate and opposed, but in which the complementarity of diverse elements is also recognized as a value in both social and ecological communities, and in which balance, harmony or optimization are the ideals.

For the way in which dualities are understood in ideological systems seems to serve as a central relational symbol for what *difference* implies. And throughout the present work it has been relational symbols or metaphors which have served as the central interpretive devices for comparing the philosophies considered.

In sum, where the dominant relational symbolism of the West is based upon opposition and separation, those of Native America and eco-holism tend to be based upon balance and complementarity. Where the former implies a separation into superior and inferior, and the pyramidal structure of social hierarchies, the latter imply egalitarianism, and thus the series of nested spheres which I have described as a holarchy. Where the former is modeled after social patterns of relationship, the latter is modeled after natural patterns of relationship; or the way in which parts and wholes are related to one another to form a larger "unity-in-diversity" in complex organic systems. Thus, the most basic difference between the two is that the relational symbolism of the West is based upon the fundamental *separateness* of all things, while those of organic philosophies are based upon the fundamental *relatedness* of all things—whether this be expressed in terms of kinship or ecology.

As was emphasized above, one of the most fundamental ways in which this bias towards separation is expressed in the dominant Western philosophy is through the many dualisms which have always characterized Western thought—whether between mind and body, culture and nature, facts and values, or science and religion. Ironically, as Deloria observes, "American Indian tribal religions...appear to be more at home in the modern world than Christian ideas and Westerner's traditional religious concepts" (1994:95). After all, in being based upon natural archetypes, their religions have always looked to both dreams and empirically observable phenomena–rather than sacred texts–in order to discover their spirituality. Thus, there is no apparent inconsistency between the religious insights of Native traditions and the findings of science, and there are no disputes equivalent to those between creationists and evolutionists. After all, both "creation" and "the sacred" may be observed in the functioning of organic systems in the present day.

Deloria's point again illustrates the relevance of Native American traditions to the broader Western society, for such a participatory philosophy provides an avenue for bridging the "gap" between our religious and scientific, or our moral and practical views of the world. In other words, unlike Occidental religions (whether Christian or economic), such a participatory philosophy could provide a model for a moral and spiritual sensibility which is consistent with the findings of contemporary science, and help to "lay the groundwork for bringing our view of the world back to a unified whole" (Deloria 1994:93). Once again, this is precisely the goal of eco-holism, which it expresses through its rejection of both mechanism and materialism, and idealism or spiritualism, as well as the duality between moral and scientific questions. For in the contemporary context, this dualism has reduced science's empirical world view to a mere servant of the dominant, inherently destructive economic ethos. Thus, there is much of a practical nature to be learned from the non-dualistic philosophies of Native America.

Yet the Western world's "separation" from nature is also much more than just a tenet of its philosophy, for it has been embodied in its practices and styles of life as well, especially through the inherently anthropocentric nature of economic thought. In practical terms, we are also separated from the direct acquisition of almost every necessity of life except the air we breath (and even this is tainted). Concerning every other fundamental necessity–whether water, food or energy–we are increasingly separated from the source of these necessities by a seemingly interminable series of intermediaries. And as Naess observes, "[i]f a great deal of technique and apparatus are placed between oneself and nature, nature cannot possibly be reached" (1989:179).

In other words, if we surround ourselves with a socially constructed, artificial world–as we indeed have–it makes it more difficult to see nature and its own patterns of relationship. Consequently, because we *do* distance ourselves from nature, we are likely to find ourselves following the only patterns of relationship which we are familiar with, and intimately in contact with on a daily basis–that of our hierarchical social structure, and of the interactions of machines. After all, one cannot follow natural models, nor adapt to natural patterns, if one has no direct experience of them, and as George observed:

> Once people knew how to live in harmony, now the silence of nature reaches few. There are many who look but only some who see...Today, harmony still lives in nature, although we have less wilderness, less variety of creatures...take care, or soon our ears will strain in vain to hear the creators' song (1974:52 and 66-7).

For it is only by looking to nature that adaptive patterns of life are made possible. Perhaps it would be helpful then, particularly in this urban age, to look to those peoples who have the longest tradition of looking to nature as a model for their own philosophies and lifestyles when seeking a more adequate understanding of nature. For respect for Mother Earth was the religion of many–if not most–indigenous peoples, and that religion was lived for centuries or millennia before the Western world arrived on their shores.

Their central realization, which the Western world now has but little choice to embrace, was also beautifully and succinctly expressed by George: "We are as much alive as we keep the earth alive" (1982:52). And since the Western world has so little experience in practicing the type of respect for nature which it now requires, while indigenous views span centuries or millennia, the least Western philosophy ought to do is to include them in the conversation. For while the Western world has long considered both Native Americans and Nature itself to be its inferiors, from which there were, therefore, no moral or ethical lessons to be learned, perhaps the time has come to reassess this thinking–for there are many such lessons to be learned from both. And as the present work has attempted to illustrate, many of the assumptions which underlie the current practices of the Western world are antithetical to the development of adaptive and sustainable life-styles. So as Eastman reminds us:

> If we are of the modern type of mind, that sees in natural law a majesty and grandeur far more impressive than any solitary infraction of it could possibly be, let us not forget that, after all, science has not explained everything. We have still to face the ultimate mystery–the origin and principle of life! Here is the supreme mystery that is the essence of worship, without which there can be no religion, and in the presence of this mystery our attitude cannot be very unlike that of the natural philosopher, who beholds with awe the Divine in all creation (1980:17-8).

Or as Standing Bear expresses a similar sentiment, in explaining why, as an elder, he came to reject the Christianity which he had learned in favour of "the natural way of my forefathers:"

> after all the great religions have been preached and expounded, or have been revealed by brilliant scholars, or have been written in books and embellished in fine language with fine covers, man–all man–is still confronted with the Great Mystery (1978:258).

Indeed, as I have tried to suggest, it is only through attaining an adequate understanding of that Great Mystery which is life, and through a thorough appreciation of the implications of such a view, that we are able to grasp the essence of an ecological sensibility. For when we recognize that we are but a small part of much larger complex organic systems, which are inherently unpredictable, all certainty goes out the window. As a consequence, perhaps it would be wise to select our technologies with more care, and to give some thought to their ecological consequences *prior* to implement- ation, rather than concerning ourselves only with their ability to maximize profits. For while error may be inevitable in dealing with complex organic systems, this does not imply that we should throw up our hands in despair and do nothing. Rather, if we are to err, it is imperative that we err on the side of caution, for we must live in and with the land if we hope to live at all. Further, we must remain close enough to those fundamental living processes so that we are able to receive feedbacks on the consequences of our errors, and correct for them.

In the final analysis, however, the life process shall always remain a Great Mystery, which we may more or less adequately *describe*, or *adapt to*, but which we shall *never* fully *explain*. And perhaps it is only when we are humbled by this fact that we may begin to learn *respect*, for as Lame Deer notes, "[m]an cannot live without mystery. He has a great need of it" (1972:163).

After all, in a world in which everything is interconnected, everything is also relevant to every decision we make. And since humans are only fragile and finite beings, a perfect comprehension of the world is obviously beyond our grasp, as was suggested by Esteva and Prakash above, in their critique of global thinking. Or as Eastman makes a similar point, "all speech is of necessity feeble and imperfect" (1980:4). And a perfect understanding is hardly to be expected from imperfect tools. Thus a profound understanding of mystery and uncertainty, and of the limits of human knowledge and power, are essential to an ecological lifestyle[168].

This more uncertain stance is also a prerequisite for the type of study I have engaged in. For aside from the ecological implications, as long as we in the West remain smugly certain of the superiority of our own beliefs–whether they go by the name of Christian dogma, scientific fact, or economic "necessity"–we can never truly open ourselves to the beliefs of another culture. Consequently, even when we have recorded the words of persons such as Black Elk or Lame Deer, and included them "in the consultable record of what humanity has said," they will remain but exhibits in a museum–idle curiosities and oddities with which to amuse ourselves–as long as we remain certain of our own cultural superiority. Only when we recognize our own inescapable uncertainty can we begin to engage in a true conversation with those texts; a dialogue in which we are open to the possibility of learning *from* other cultures, and not merely *about* them. Indeed, aspects of our own ways of thinking and living may even be shown to be inadequate in the process.

In this way we may come to realize that the opinions of peoples from other cultures are more than merely data to be analyzed, and to be appropriately fitted into a catalogue of cultural types. Instead, we can respectfully include these persons among our *teachers* and *colleagues*, and treat their opinions with the same respect as we treat those of Western scholars.

This is precisely what I have attempted to do throughout the present work. For the desire to nurture and promote what we now call an "ecological lifestyle," or an "ecological sensibility," has long been at the core of Native teachings, particularly those of their

---

[168] And perhaps this is why learning from the symbolism of dreams and visions has always been so central in many Native American cultures.

shamans or holy people, and in the contemporary situation, their scholars, who fulfill an analogous role. For as Black Elk pointed out, in his day, "[t]he medicine men were the learned class, the scholars of the tribe" (1984:334).

More important, however, is the fact that Native American philosophies appear always to have attempted to *embody* such a philosophy in their patterns of life. It is not merely an abstract philosophy promulgated by a few of the most articulate men and women, nor a hegemonic ideology promoted by an elite in order to advance their own narrow interests, but rather a philosophy *of living* which was–*and still is*–practiced by human communities, however imperfectly or well.

And the sooner we learn to distinguish between economics and reality, as well as between what is in reality only in the short-term interests of the capitalist class, and that which is in the long-term interests of life on Earth, the sooner we may be able to recreate such a lived philosophy, and the sooner we may be able to create a viable future for our children. Thus, as LaDuke concludes:

> In the final analysis, the survival of Native America is fundamentally about the collective survival of human beings. The question of who gets to determine the destiny of the land, and of the people who live on it–those with the money or those who pray on the land–is a question that is alive throughout society (1999:5).

Hetch etu aloh. Ina Maka awanyanka.

Bibliography

Agrawal, Arun
(1995) "Dismantling the Divide Between Indigenous and Scientific
Knowledge." *Development and Change* 26:413-39.
Aquinas, Saint Thomas
(1945) "On Truth," in: *Basic Writings of Saint Thomas Aquinas, Vol. 1,*
edited by Anton C. Pegis, New York: Random House.
Aristotle
(1986) *Selected Works,* translated by Hippocrates G. Apostle and Lloyd P.
Gerson, Grinnel, Iowa: The Peripatetic Press.
Augustine, Saint
(1853) "On the Free Will," in: *Augustine: Earlier Writings,* Vol. VI,
translated by John H. S. Burleigh, Philadelphia: The Westminster Press.
Barker, Ernest
(1960) *Social Contract: Essays by Locke, Hume and Rousseau.* London:
Oxford University Press.
Bates, Daniel G. and Elliot M. Fratkin
(1999) *Cultural Anthropology,* 2nd ed. Toronto: Allyn and Bacon.
Bateson, Gregory
(1972) *Steps to an Ecology of Mind.* New York: Ballantine Books.
(1979) *Mind and Nature: A Necessary Unity.* New York: Bantam Books.
Bateson, Gregory and Mary Catherine Bateson
(1988) *Angels Fear: Towards an Epistemology of the Sacred.* New York:
Bantam Books.
Baxter, William
(1986) "People or Penguins," in: *People, Penguins and Plastic Trees: Basic
Issues in Environmental Ethics,* edited by VanDeVeer and Pierce, Belmont,
Calif.: Wadsworth Publishing Co., pp. 214-18.
Berger, Thomas R.
(1988) *Northern Frontier Northern Homeland: The Report of the
MacKenzie Valley Pipeline Inquiry,* revised edition. Toronto: Douglas and
McIntyre.
Berkes, Fikret
(1998) "Indigenous Knowledge and Resource Management Systems in the
Canadian Subarctic," in: *Linking Social and Ecological Systems:
Management Practices and Social Mechanisms for Building Resilience,*
edited by Berkes and Folke, Cambridge: Cambridge University Press, pp.
98-128.
(1999) *Sacred Ecology: Traditional Ecological Knowledge and Resource
Management..* Philadelphia: Taylor and Francis.
Berkes, Fikret, Johan Colding and Carl Folke
(2003) "Introduction," in: *Navigating Social-Ecological Systems: Building
Resilience for Complexity and Change,* edited by Berkes, Colding and
Folke, Cambridge: Cambridge University Press, pp. 1-29.
Berkes, Fikret and H. Fast
(1996) "Aboriginal Peoples: The Basis for Policy-making Towards

Sustainable Development," in: *Achieving Sustainable Development*, edited by Dale and Robinson, Vancouver: University of British Columbia Press, pp. 204-64.

Berkes, Fikret and Carl Folke
(1998) "Linking Social and Ecological Systems for Resilience and Sustainability," in: *Linking Social and Ecological Systems: Management Practices and Social Mechanisms for Building Resilience.* edited by Berkes and Folke, Cambridge: Cambridge University Press, pp. 1-25.

Berkes, Fikret, Carl Folke and Madhav Gadgil
(1995) "Traditional Ecological Knowledge, Biodiversity, Resilience and Sustainability," in: *Biodiversity Conservation*, edited by Perrings et al., Netherlands: Kluwer Academic Publishers, pp. 281-99.

Berkes, Fikret, P. J. George, R. J. Preston, A. Hughes, J. Turner, B. D. Cummins
(1994) "Wildlife Harvesting and Sustainable Regional Native Economy in the Hudson and James Bay Lowland, Ontario." *Arctic* 47(4):350-69.

Berkes, Fikret and Alison Haugh
(1992) "Wildlife Harvests in the Moose River Basin." Report prepared for the Moose River/James Bay Coalition.

Berkes, Fikret and Thomas Henley
(1997) "Co-management and Traditional Knowledge: Threat or Opportunity?" *Policy Options*, March, pp. 29-31.

Berkes, Fikret, A. Hughes, P. J. George, R. J. Preston, B. D. Cummins, J. Turner
(1995) "The Persistence of Aboriginal Land Use: Fish and Wildlife Harvest Areas in the Hudson and James Bay Lowland, Ontario." *Arctic* 48(1):81-93.

Berkes, Fikret, Mina Kislaioglu, Carl Folke and Madhav Gadgil
(1998) "Exploring the Basic Ecological Unit: Eco-system-like Concepts in Traditional Societies." *Ecosystems* 1(5):1-7.

Berkhofer, Jr., Robert F.
(1978) *The White Man's Indian: Images of the American Indian from Columbus to the Present.* New York: Vintage Books.

Berman, Morris
(1984) *The Reenchantment of the World.* New York: Bantam Books.

Bernstein, Henry
(2000) "Colonialism, Capitalism and Development," in: *Poverty and Development into the 21st Century,* edited by Allen and Thomas, Oxford: Oxford University Press, pp. 241-70.

Berry, Wendel
(1977) *The Unsettling of America: Culture and Agriculture.* San Francisco: Sierra Club Books.
(1981) *The Gift of Good Land: Further Essays Cultural and Agricultural.* San Francisco: North Point Press.

Berwick, Dennison
(1992) *Savages: The Life and Killing of the Yanomami.* Toronto: MacFarlane Walter and Ross.

Biehl, Janet
(1991) *Finding Our Way: Rethinking Ecofeminist Politics*. Montreal: Black
Rose Books.
Bird-David, Nurit
(1993) "Tribal Metaphorization of Human-Nature Relatedness: A
Comparative Analysis." In: *Environmentalism: The View from
Anthropology*, edited by Kay Milton. London: Routledge, pp.112-25.
Black Elk, with Joseph Epes Brown, ed.
(1971) *The Sacred Pipe: Black Elks Account of the Seven Rites of the
Oglala Sioux*. New York: Viking/Penguin. First published in 1953.
Black Elk, with John G. Neihardt, ed.
(1988) *Black Elk Speaks: Being the Life Story of a Holy Man of the Oglala
Sioux*. Lincoln: University of Nebraska Press. First published in 1932.
Bodley, John H.
(1999) *Victims of Progress*. 4th ed. Toronto: Mayfield Publishing Company.
(2001) *Anthropology and Contemporary Human Problems*, 4th ed. Toronto:
Mayfield Publishing Company.
Bookchin, Murray
(1980) *Toward an Ecological Society*. Montreal: Black Rose Books.
(1986) *Post-Scarcity Anarchism*, 2nd ed., Montreal: Black Rose Books.
(1990)*The Philosophy of Social Ecology: Essays on Dialectical
Naturalism*. Montreal: Black Rose Books.
(1991)*The Ecology of Freedom: The Emergence and Dissolution of
Hierarchy.*, 2nd ed., Montreal: Black Rose Books.
Booth, Annie L. and Harvey M. Jacobs
(1990) "Ties That Bind: Native American Beliefs as a Foundation for
Environmental Consciousness." *Environmental Ethics* 12, pp. 27-43.
Bright, Christopher
(1999) "Invasive Species: Pathogens of Globalization." *Foreign Policy*,
Fall, pp. 50-64.
(2001) "The Nemesis Effect," in: *Environment 01/02.*, edited by Allen.
Guilford, Conn.: McGraw-Hill/Dushkin, pp. 14-23.
Brightman, Robert A.
(1993) *Grateful Prey: Rock Cree Human-Animal Relationships*. Berkeley:
University of California Press.
Brouwer, Jan
(1998) "On Indigenous Knowledge and Development." *Current
Anthropology* 39(3):351.
Brown, Dee
(1970) *Bury My Heart at Wounded Knee: An Indian History of the
American West*. New York: Henry Holt and Co.
Burger, Joanna and Michael Gochfeld
(2000) "The Tragedy of the Commons: Thirty Years Later," in: *Environment
00/01*, edited by Allen. Guilford, Conn.: McGraw-Hill/Dushkin, pp. 126-33.

Cajete, Gregory
(2000) *Native Science: Natural Laws of Interdependence*. Santa Fe: Clear Light Publishers.
Callicott, J. Baird
(1982) "Traditional American Indian and Western European Attitudes Towards Nature." *Environmental Ethics* 4:293-318.
(1989) *In Defense of the Land Ethic: Essays in Environmental Philosophy*. Albany: State University of New York Press.
(1994) *Earths Insights: A Survey of Ecological Ethics from the Mediterranean Basin to the Australian Outback*. Berkeley:University of California Press.
Calvin, William H.
(2000) "The Great Climate Flip-flop," in: *Environment 00/01*, edited by Allen, Guilford, Conn.: McGraw-Hill/Dushkin, pp. 167-74.
Campbell, Colin J. and Jean H. LaHerrere
(1998) "The End of Cheap Oil." *Scientific American*, March, pp. 78-83.
Campbell, Joseph
(1964) *The Masks of God: Occidental Mythology*. New York: Penguin.
(1969) *The Masks of God: Primitive Mythology*, revised ed., New York: Penguin.
(1984) "Joseph Campbell on the Great Goddess," *Parabola* 4, pp. 75-76.
Capra, Fritjof
(1975) *The Tao Of Physics: An Exploration of the Parallels Between Modern Physics and Eastern Mysticism*. London: Fontana.
(1982) *The Turning Point: Science, Society, and the Rising Culture*. New York: Bantam Books.
Cardinal, Harold
(1969) *The Unjust Society: The Tragedy of Canadas Indians*. Edmonton: M. G. Hurtig.
Carson, Rachel
(1962) *Silent Spring*. Boston: Houghton Mifflin Co.
Chodkiewicz, Jean-Luc and Jennifer Brown
(1999) *First Nations and Hydroelectric Development in Northern Manitoba: The Northern Flood Agreement: Issues and Implications*. Winnipeg: Centre for Rupertsland Studies, University of Winnipeg.
Clark, John
(1984) *The Anarchist Moment: Reflections on Culture, Nature and Power*, Montreal: Black Rose Books.
(1992) "What is Social Ecology?" *Our Generation* 23(1):91-98.
Clifton, James A., ed.
(1990) *The Invented Indian: Cultural Fictions and Government Policies*. New Brunswick(USA): Transaction Publishers.
Cordova, V. F.
(1997) "Ecoindian: A Response to J. Baird Callicott." *Ayaangwaamizin: International Journal of Indigenous Philosophy*. 1(1):31-44.

Cree Trappers Association
(1989) *Cree Trappers Speak*. Chisasibi: James Bay Cree Cultural Education Centre.

Cronon, William
(1983) *Changes in the Land: Indians, Colonists and the Ecology of New England*. New York: Hill and Wang.

Daly, Herman E.
(1995) *The Steady-State Economy*, 2nd ed., Washington, D. C.: Island Press.

Daly, Herman E. and John B. Cobb
(1989) *For the Common Good: Redirecting the Economy Toward Community, the Environment, and a Sustainable Future*. Boston: Beacon Press.

Dasmann, R. F.
(1988) "Towards a Biosphere Consciousness," in: *The Ends of the Earth*, edited by Worster, Cambridge: Cambridge University Press.

Deloria, Jr., Vine
(1970) *Custer Died For Your Sins: An Indian Manifesto*. Norman: University of Oklahoma Press.
(1994) *God is Red: A Native View of Religion*, 2nd ed., Golden, Colorado: Fulcrum Publishing.

DeMallie, Raymond J., ed.
(1985) *The Sixth Grandfather: Black Elks Teachings Given to John G. Neihardt* . Lincoln. University of Nebraska Press.

Descartes, Rene
(1980) *Discourse on the Method of Rightly Conducting Ones Reason and For Seeking Truth in the Sciences*, translated by Donald A. Cress, Indianapolis:Hackett Publishing Co. Originally published in 1637.
(1988) *Selected Philosophical Works*, trans. by John Cottingham, Robert Stoothoff and Dugald Murdoch, Cambridge: Cambridge University Press.

Desmore, Francis
(1918) *Teton Sioux Music*, Washington, D. C.: Bulletin 61, Bureau of Ethnography.

Devall, Bill
(1980) "The Deep Ecology Movement." *Natural Resources Journal* 20:299-322.

Devall, Bill and George Sessions
(1985) *Deep Ecology: Living As If Nature Mattered*. Salt Lake City: Peregrine Smith Books.

Diamond, Irene and Gloria Feman Orenstein, eds.
(1990) *Reweaving the World: The Emergence of Ecofeminism*. San Francisco: Sierra Club Books.

Dion, Joseph
(1979) *My Tribe the Crees*. Calgary: Glenbow.

Donald, Leland
(1990) "Liberty, Equality, Fraternity: Was the Indian Really Egalitarian?" in: *The Invented Indian: Cultural Fictions and Government*

Policies, edited by Clifton, New Brunswick(USA): Transaction Publishers, pp. 145-67.

Dudgeon, Roy C.
(2008) *The Pattern Which Connects: An Eco-holist Critique of Postmodernism.* Winnipeg: Pitch Black Publications.

Dudgeon, Roy C. and Fikret Berkes
(2003) "Local Understandings of the Land: Traditional Ecological Knowledge and Indigenous Knowledge," in: *Nature Across Cultures: Views of Nature and the Environment in Non-Western Cultures*, edited by H. Selin, Lancaster, U. K.: Kluwer Academic Publishers, pp. 75-96.

Dunn, Seth
(1999) "King Coal's Weakening Grip on Power," *World Watch*, September/October, pp. 10-19.
(2001) "The Hydrogen Experiment," in: *Environment 01/02.*, edited by Allen, Guilford, Conn.: McGraw-Hill/Dushkin, pp. 87-96.

Easterbrook, Gregg
(1999) "Warming Up: The Real Evidence for the Greenhouse Effect," *The New Republic*, November 8, pp. 42-53.

Eastman, Charles A.
(1971) *Indian Boyhood.* New York: Dover Publications. First published in 1902.
(1977) *From Deep Woods to Civilization: Chapters in the Autobiography of an Indian.* Lincoln: University of Nebraska Press. First published in 1916.
(1980)*The Soul of the Indian: An Interpretation.* Lincoln: University of Nebraska Press. First published in 1911.

Eliade, Mircea
(1954) *The Myth of the Eternal Return or, Cosmos and History.* Princeton: Princeton University Press.

Erdoes, Richard and Alfonso Ortiz, eds.
(1984) *American Indian Myths and Legends.* New York: Pantheon Books.

Esteva, Gustava and Mahdu Suri Prakash
(1997) "From Global Thinking to Local Thinking," in: *The Post-development Reader.*, edited by Kahnema and Bawtree, London: Zed Books, pp. 277-89.

Feit, Harvey A.
(1973) "Ethno-ecology of the Waswanipi Cree; or How Hunters Can Manage Their Resources," in: *Cultural Ecology*, edited by Cox, Toronto: McClelland and Stewart, pp. 115-25.
(1979) "Political Articulations of Hunter to the State: Means of Resisting Threats to Subsistence Production in the James Bay and Northern Quebec Agreement." *Inuit Studies* 3(2):37-52.

Fenge, Terry
(1997) "Ecological Change in the Hudson Bay Bioregion: A Traditional Ecological Knowledge Perspective." *Northern Perspectives* 25(1):2-3.

Flavin, Christopher
(2000) "Last Tango in Buenos Aires," in: *Environment 00/01*, edited by Allen, Guilford, Conn.: McGraw-Hill/Dushkin, pp. 175-81.
(2001) "Bull Market in Wind Energy," in: *Environment 01/02.*, edited by Allen, Guilford, Conn.: McGraw-Hill/Dushkin, pp. 103-5.

Fools Crow, Frank, with Thomas E. Mails, ed.
(1979) *Fools Crow*. Lincoln: University of Nebraska Press.
(1991) *Fools Crow: Wisdom and Power*. Tulsa, Oklahoma: Council Oak Books.

Fox, Warwick
(1986) *Approaching Deep Ecology: A Response to Richard Sylvan's Critique*. Hobart: University of Tasmania, Centre for Environmental Studies.
(1990) *Towards a Transpersonal Ecology: Developing New Foundations for Environmentalism*. Boston: Shambhala.

Francis, Daniel
(1992) *The Imaginary Indian: The Image of the Indian in Canadian Culture*. Vancouver: Arsenal Pulp Press.

Gadgil, M.
(1987) "Diversity: Cultural and Biological." *Trends in Ecology and Evolution* 2:369-73.

Gadgil, M. Fikret Berkes and Carl Folke
(1993) "Indigenous Knowledge for Biodiversity Conservation." *Ambio* 22: 151-6.

Gedicks, Al
(1993) *The New Resource Wars: Native and Environmental Struggles Against Multinational Corporations*. Boston: South End Press.
(2001) *Resource Rebels: Native Challenges to Mining and Oil Corporations*. Boston: South End Press.

Geertz, Clifford
(1973) *The Interpretation of Cultures*. New York: Basic Books.

George, Chief Dan, with Helmut Hirnschall, ed.
(1974) *My Heart Soars*. Surrey, B. C.: Hancock House.
(1982) *My Spirit Soars*, Surrey, B. C.: Hancock House.

George, Peter, Fikret Berkes and Richard J. Preston
(1995) "Aboriginal Harvesting in the Moose River Basin: A Historical and Contemporary Analysis." *The Canadian Review of Sociology and Anthropology* 32(1):69-90.

Gifford, Eli and Michael Cook, eds.
(1992) *How Can One Sell the Air? Chief Seattle's Vision*. Summertown, Tenn.: The Book Publishing Co.

Gill, Sam
(1987) *Mother Earth: An American Story*. Chicago: University of Chicago Press.
(1990) "Mother Earth: An American Myth." In: *The Invented Indian:*

Cultural Fictions and Government Policies. edited by Clifton. New Brunswick(USA): Transaction Publishers, pp. 129-43.

Glavin, Terry
(1990) *A Death Feast in Dimlahamid.* Vancouver: New Star Books.

Goddard, John
(1991) *Last Stand of the Lubicon Cree.* Toronto: Douglas and McIntyre.

Goldschmidt, Walter
(1978) *As You Sow: Three Studies in the Social Consequences of Agribusiness.* Montclair, N. J.: Allanheld, Osmun and Co.

Goldstick, Miles
(1987) *Wollaston: People Resisting Genocide.* Montreal: Black Rose Books.

Gould, Stephen J. and N. Eldredge
(1977) "Punctuated Equilibria: The Tempo and Mode of Evolution Reconsidered." *Paleobiology* 3:115-51.

Griffin, David Ray, ed.
(1988) *The Reenchantment of Science: Postmodern Proposals.* Albany: State University of New York Press.

Griffin, Susan
(1978) *Woman and Nature: The Roaring Inside Her.* New York: Harper and Row.

Gunderson, L. H., C. S. Holling and S. S. Light, eds.
(1995) *Barriers and Bridges to the Renewal of Ecosystems and Institutions.* New York: Columbia University Press.

Halifax, Joan
(1979) *Shamanic Voices: A Survey of Visionary Narratives.* New York: E. P. Dutton.

Halweil, Brian
(2001) "Where Have All the Farmers Gone?," in: Allen, ed., *Environment 01/02.* Guilford, Conn.: McGraw-Hill/Dushkin, pp. 250-61.

Hardin, Garret
(1968) "The Tragedy of the Commons," *Science* 162:1243-8.

Harries-Jones, Peter
(1992) "Sustainable Anthropology: Ecology and Anthropology in the Future," in: *Contemporary Futures: Perspectives From Social Anthropology,* edited by Wallman, London: Routledge.
(1995) *A Recursive Vision: Ecological Understanding and Gregory Bateson.* Toronto: University of Toronto Press.

Harris, Marvin
(1979) *Cultural Materialism: The Struggle for a Science of Culture.* New York: Vintage Books.

Harrod, Howard L.
(2000) *The Animals Came Dancing: Native American Sacred Ecology and Animal Kinship.* Tucson: University of Arizona Press.

Hefernden, James D.
(1982) "The Land Ethic: A Critical Appraisal." *Environmental Ethics*

4:235-47.

Hester, Thomas Lee and Dennis McPherson
(1997) "The Euro-American Philosophical Tradition and its Ability to
Examine Indigenous Philosophy." *Ayaangwaamizin: International Journal
of Indigenous Philosophy* 1(1):3-9.

Hobbes, Thomas
(1985) *Leviathan*. New York: Penguin. First published in 1651.

Hoebel, E. Adamson
(1978) *The Cheyennes: Indians of the Great Plains*, 2nd ed., New York:
Holt, Rinehart and Winston.

Holling, C. S.
(1973) "Resilience and Stability of Ecological Systems." *Annual Review of
Ecology and Systematics* 4:1-23.

Holling, C. S., ed.
(1978) *Adaptive Environmental Assessment and Management*. London:
Wiley.

Holling, C. S., Fikret Berkes and Carl Folke
(1998) "Science, Sustainability and Resource Management," in: *Linking
Social and Ecological Systems: Management Practices and Social
Mechanisms for Building Resilience,* edited by Berkes and Folke,
Cambridge: Cambridge University Press, pp. 342-62.

Holling, C. S. and S. Sanderson
(1996) "Dynamics of (Dis)harmony in Ecological and Social Systems," in:
Rights to Nature, edited by Hanna, Folke and Maler, Washington, D. C.:
Island Press, pp. 57-85.

Howard, Albert and Francis Widdowson
(1996) "Traditional Knowledge Threatens Environmental Assessment."
Policy Options, November, pp. 34-6.
(1997) "Revisiting Traditional Knowledge." *Policy Options*, April, pp. 46-
8.

Hosansky, David
(2001) "Mass Extinction," in: *Environment 01/02.*, edited by Allen. Guilford,
Conn.: McGraw-Hill/Dushkin, pp. 123-9.

Hughes, J. Donald
(1983) *American Indian Ecology*. El Paso: Texas Western Press.
(1991) "Metakuyase." *The Trumpeter*. 8(4):184-5.

Hume, David
(1962) "An Inquiry Concerning Human Understanding," in: *On Human
Nature and the Understanding,* edited by Flew, New York: Collier. First
published in 1777.

Ingold, Tim
(1990) "An Anthropologist Looks At Biology." *Man* (N. S.) 25, pp. 208-209.
(1992) "Editorial." *Man* (N. S.) 27, pp. 693-696.

Johannes, Robert E., ed.
(1989) *Traditional Ecological Knowledge: A Collection of Essays*. Gland:

International Conservation Union.

Johnston, Basil H.
(1982) *Ojibway Ceremonies*. Toronto: McClelland and Stewart.

Karl, Thomas R. and Kevin E. Trenberth
(1999) "The Human Impact on Climate," *Scientific American*, December, pp. 100-5.

Kates, Robert W.
(2001) "Population and Consumption: What We Know, What We Need to Know," in: *Environment 01/02.*, edited by Allen. Guilford, Conn.: McGraw-Hill/Dushkin, pp. 44-51.

Knudtson, Peter and David Suzuki
(1992) *The Wisdom of the Elders*. New York: Random House.

Koestler, Arthur
(1978) *Janus: A Summing Up*. New York: Random House.

Kohr, Leopold
(1978) *The Breakdown of Nations*. New York: Dutton. First published in 1957.

Krech III, Shepard
(1999) *The Ecological Indian: Myth and History*. New York: W. W. Norton and Co.

Krech III, Shepard, ed.
(1981) *Indians, Animals and the Fur Trade: A Critique of KEEPERS OF THE GAME*. Athens: University of Georgia Press.

Kuhn, Thomas
(1970) *The Structure of Scientific Revolutions*, 2nd ed., Chicago: University of Chicago Press.

LaDuke, Winona
(1999) *All Our Relations: Native Struggles for Land and Life*. Cambridge: South End Press.

Lame Deer, John (Fire), with Richard Erdoes, ed.
(1972) *Lame Deer: Seeker of Visions*. New York: Washington Square Press.

Lashof, Daniel A.
(2000) "The Earth's Last Gasp?" in: *Environment 00/01*, edited by Allen, Guilford, Conn.: McGraw-Hill/Dushkin, pp. 212-14.

Leibniz, G. W.
(1989) *Philosophical Essays*, translated by Roger Ariew and Daniel Garber, Indianapolis: Hackett Publishing. Co.

Leopold, Aldo
(1949) *A Sand County Almanac: And Sketches Here and There*. Oxford: Oxford University Press.

Lewellen, Ted C.
(1992) *Political Anthropology: An Introduction*, 2nd ed. London: Bergin and Garvey.

Lewin, Roger
(1992) *Complexity: Life at the Edge of Chaos*. New York: Collier Books.

Locke, John
(1960) "An Essay Concerning the True Original, Extent and End of Civil Government." In: *Social Contract*. edited by Barker, Oxford: Oxford University Press, pp. 1-144. First published in 1690.
Lovelock, J. E.
(1979) *Gaia: A New Look at Life on Earth*. Oxford: Oxford University Press.
Martin, Calvin
(1978) *Keepers of the Game: Indian-Animal Relationships and the Fur Trade*. Berkeley: Berkeley University Press.
Maybury-Lewis, David
(2002) *Indigenous Peoples, Ethnic Groups and the State*, 2nd ed., Boston: Allyn and Bacon.
McCreight, M. I.
(1947) *Firewater and Forked Tongues, A Sioux Chief Interprets U. S. History*. Pasadena: Trails End Publishing Co.
McGaa, Ed (Eagle Man)
(1990) *Mother Earth Spirituality: Native American Paths to Healing Ourselves and Our World*. San Francisco: Harper.
McGinn, Anne Platt
(2000) "POPs Culture," *World Watch*, March/April, pp. 26-36.
McGrew, Anthony
(2000) "Sustainable Globalization? The Global Politics of Development and Exclusion in the New World Order." in: *Poverty and Development into the 21st Century,*, edited by Allen and Thomas, Oxford: Oxford University Press, pp. 345-64.
McLuhan, T. C., ed.
(1971) *Touch the Earth: A Self-Portait of Indian Existence*. New York: Touchstone Press.
McNeil, William H.
(1976) *Plagues and Peoples*. Garden City, N. Y.: Anchor Books.
McNickle, DArcy
(1973) *Native American Tribalism: Indian Survivals and Renewals*. Oxford: Oxford University Press.
McPherson, Dennis H. and J. Douglas Rabb
(1993) *Indian From the Inside: A Study in Ethno-Metaphysics*. Thunder Bay, Ont.: Centre for Northern Studies, Lakehead University.
Meadows, Donella, Dennis Meadows and Jorgan Randers
(1992) *Beyond the Limits: Confronting Global Collapse, Envisioning a Sustainable Future*, Post Mills, Vt.: Chelsea Green Publishing Co.
Meadows, Donella, Dennis Meadows, Jorgan Randers and William Behrens III
(1972) *The Limits to Growth*. New York: Universe.
Merchant, Carolyn
(1980) *The Death of Nature: Women, Ecology and the Scientific Revolution*. San Francisco: HarperCollins.

(1992) *Radical Ecology: The Search for a Liveable World*. London: Routledge.

Miller, Peter
(1983) "Do Animals Have Interests Worth of Our Moral Interest?," *Environmental Ethics* 5:319-33.

Momaday, N. Scott
(1976) "A First American Views His Land," *National Geographic* 150(1):13-18.

Morantz, Toby
(1986) "Historical Perspectives on Family Hunting Territories in Eastern James Bay." *Anthropologica* 28(1-2):65-91.

Morrison, James
(1992) "Colonization, Resource Extraction and Hydroelectric Development in the Moose River Basin: A Preliminary History of the Implications for Aboriginal People." Report prepared for the Moose River/James Bay Coalition.

Nabokov, Peter
(1991) *Native American Testimony: A Chronicle of Indian-White Relations from Prophecy to Present, 1492-1992*. New York: Viking/Penguin.

Naess, Arne
(1973) "The Shallow and the Deep, Long-Range Ecology Movement." *Inquiry* 16:95-100.
(1989) *Ecology, Community and Lifestyle: Outline of an Ecosophy*, translated and edited by Rothenberg. Cambridge: Cambridge University Press.

National Film Board of Canada
(1990) *The Spirit Within*.

Niezen, Ronald
(1998) *Defending the Land: Sovereignty and Forest Life in James Bay Cree Society*. Toronto: Allyn and Bacon.
(1999) "Treaty Violations and the Hydro-Repayment Rebellion of Cross Lake, Manitoba." *Cultural Survival*, Spring, pp. 18-21.

Nikiforuk, Andrew
(1996) *The Fourth Horseman*. Toronto: University of Toronto Press.

Odum, Eugene
(1989) *Ecology and Our Endangered Life-Support System*. Sunderland, Mass.: Sinauer Associates.

Ong, Aihwa
(1987) *Spirits of Resistance and Capitalist Discipline: Factory Women in Malaysia*, Albany: State University of New York Press.

Plato
(1974) *The Republic*, translated by Lee, New York: Penguin.

Porritt, Jonathon and David Winner
(1988) *The Coming of the Greens*. London: Fontana.

Posey, Darrel Addison
(1998) "Comment on 'The Development of Indigenous Knowledge: A New Applied Anthropology' by Paul Sillitoe." *Current Anthropology* 39(2):241-2.

Potter, David
(2000) "The Power of Colonial States," in: *Poverty and Development into the 21st Century*, edited by Allen and Thomas, Oxford: Oxford University Press, pp. 271-88.

Preston, Richard J.
(2002) *Cree Narrative: Expressing the Personal Meaning of Events*, 2nd edition. Montreal: McGill-Queen's University Press.

Preston, Richard J., Fikret Berkes and Peter J. George
(1995) "Perspectives on Sustainable Development in the Moose River" Basin, in: *Papers of the Twenty-Sixth Algonquian Conference*, edited by David H. Pentland, Winnipeg: University of Manitoba Press, pp. 379-94.

Pretty Shield, with Frank B. Linderman, ed.
(1974) *Pretty Shield: Medicine Woman of the Crows*. Lincoln: University of Nebraska Press. First published in 1932.

Price, T. Douglas and Gary M. Feinman
(1993) *Images of the Past*. Mountain View, Ca.: Mayfield Publishing Co.

Purcell, Trevor W.
(1998) "Indigenous Knowledge and Applied Anthropology: Questions of Definition and Development." *Human Organization* 57(3):258-72.

Quammen, David
(2001) "Planet of Weeds: Tallying the Losses of Earth's Animals and Plants," in: *Environment 01/02.*, edited by Allen, Guilford, Conn.: McGraw-Hill/Dushkin, pp. 108-15.

Quinn, Randy
(2000) "Sunlight Brightens Our Future," in: *Environment 00/01*, edited by Allen, Guilford, Conn.: McGraw-Hill/Dushkin, pp. 80-4.

Rappaport, Roy A.
(1979) *Ecology, Meaning and Religion*. Berkeley: North Atlantic Books.
(1994) "Humanity's Evolution Anthropologys Future," in: *Assessing Cultural Anthropology*, edited by Borofsky, New York: McGraw-Hill, pp. 153-66.

Ray, Arthur J.
(1974) *Indians in the Fur Trade: Their Role as Hunters, Trappers and Middlemen in the Lands Southwest of Hudsons Bay, 1660-1870*. Toronto: University of Toronto Press.

Ray, Carl and James Stevens
(1971) *Sacred Legends of the Sandy Lake Cree*. Toronto: McClelland and Stewart.

RCAP (Royal Commission on Aboriginal Peoples)
(1997) *For Seven Generations: An Information Legacy of the Royal Commission on Aboriginal Peoples*. Ottawa: Libraxus, CD-ROM.

Reed, Gerard
(1991) "A Sioux View of the Land: The Environmental Perspectives of Charles A. Eastman." *The Trumpeter.* 8(4):170-73.
Rice, Julian
(1991) *Black Elks Story: Distinquishing Its Lakota Purpose.* Albuquerque: University of New Mexico Press.
Regan, Tom
(1986) "The Case for Animal Rights" and "The Rights View," in: *People, Penguins and Plastic Trees: Basic Issues in Environmental Ethics,* edited by VanDeVeer and Pierce, Belmont, Calif.: Wadsworth Publishing Co., pp. 32-29 and 203-5.
Robbins, Richard H.
(1999) *Global Problems and the Culture of Capitalism.* Toronto: Allyn and Bacon.
(2002) *Global Problems and the Culture of Capitalism,* 2nd ed. Toronto: Allyn and Bacon.
Robinson, Angela
(2005) *Ta'n Teli-kltamsitasit (Ways of Believing).* Toronto: Pearson Education Canada.
Rodman, John
(1986) "Ecological Sensibility," in: *People, Penguins and Plastic Trees: Basic Issues in Environmental Ethics,* edited by VanDeVeer and Pierce, Belmont, Calif.: Wadsworth Publishing Co., pp. 165-8.
Rogers, Raymond A.
(1995) *The Oceans are Emptying: Fish Wars and Sustainability.* Montreal: Black Rose Books.
Rolston III, Holmes
(1988) *Environmental Ethics: Duties to and Values in the Natural World.* Philadelphia: Temple University Press.
Roseberry, William
(1988) "Political Economy." *Annual Review of Anthropology.* 17:161-85.
Rosentreter, Richard
(2001) "Oil, Profit$, and the Question of Alternative Energy," in: *Environment 01/02.,* edited by Allen, Guilford, Conn.: McGraw-Hill/Dushkin, pp. 79-82.
Rostow, W. W.
(1960) *The Stages of Economic Growth.* Cambridge: Cambridge University Press.
Rousseau, Jean-Jacques
(1984) *Discourse on Inequality.* New York: Penguin. First published in 1755.
Ryle, Gilbert
(1949) *The Concept of Mind.* New York: Penguin Books.
Sahlins, Marshall
(1972) *Stone Age Economics.* Chicago: Aldine Publishing Company.

Schumacher, E. F.
    (1973) *Small is Beautiful: Economics as if People Mattered.* New York: Harper and Row.
    (1977) *A Guide for the Perplexed.* New York: Harper and Row.
    (1979) *Good Work.* New York: Harper and Row.
    (1997)*This I Believe: And Other Essays.* Devon, England: Green Books Ltd.
Scott, Colin
    (1996) "Science for the West, Myth for the Rest?: The Case of James Bay Cree Knowledge Construction," in: *Naked Science*, edited by Nader, New York: Routledge.
Scott, James C.
    (1985) *Weapons of the Weak: Everyday Forms of Peasant Resistance.* New Haven: Yale University Press.
Sen, Amartya
    (1981) *Poverty and Famines: An Essay on Entitlement and Deprivation.* Oxford: Clarendon Press.
Sillitoe, Paul
    (1998) "The Development of Indigenous Knowledge: A New Applied Anthropology." *Current Anthropology*, 39(2):223-52.
Singer, Peter
    (1986) "Animal Liberation," in: *People, Penguins and Plastic Trees: Basic Issues in Environmental Ethics*, edited by VanDeVeer and Pierce, Belmont, Calif.: Wadsworth Publishing Co., pp. 24-32.
Singh, Narindar
    (1976) *Economics and the Crisis of Ecology*, 3rd edition. London: Bellew Publishing.
Sioui, Georges E.
    (1992) *For an Amerindian Autohistory: An Essay on the Foundations of a Social Ethic*, translated by Fischman. Montreal: McGill-Queens University Press.
Sklair, Leslie
    (1991) *Sociology of the Global System.* Baltimore: Johns Hopkins University Press.
Sole, R. V. and S. A. Levin, eds.
    (2002) "The Biosphere as a Complex Adaptive System: Papers of a Theme Issue," *Philosophical Transactions of the Royal Society: Biological Sciences*, Vol. 357, No. 1421, (29 May 2002), pp. 615-725.
Solomon, Arthur
    (1994) *Eating Bitterness: A Vision Beyond the Prison Walls.* Toronto: NC Press Limited.
Speck, Frank G.
    (1977) *Naskapi: The Savage Hunters of the Labrador Peninsula.* Norman: University of Oklahoma Press. First published in 1935.
Spretnak, Charlene and Fritjof Capra
    (1986) *Green Politics: The Global Promise*, 2nd edition, Santa Fe: Bear and

Company.

Standing Bear, Luther
(1975) *My People the Sioux.* Lincoln: University of Nebraska Press. First published in 1928.
(1978) *Land of the Spotted Eagle.* Lincoln: University of Nebraska Press. First published in 1933.

Stevenson, Marc G.
(1997) "Ignorance and Prejudice Threaten Environmental Assessment." *Policy Options,* March, p. 25-8.

Stirrat, R. L.
(1998) Comment on 'The Development of Indigenous Knowledge: A New Applied Anthropology' by Paul Sillitoe." *Current Anthropology* 39(2):242-3.

Stone, M. Priscilla
(1998) "Comment on The Development of Indigenous Knowledge: A New Applied Anthropology by Paul Sillitoe. *Current Anthropology* 39(2):243.

Tanner, Adrian
(1979) *Bringing Home Animals.* London: Hurst.

Taylor, Paul
(1986) *Respect for Nature: A Theory of Environmental Ethics.* Princeton: Princeton University Press.

Teeple, Gary
(2000) *Globalization and the Decline of Social Reform: Into the Twenty-first Century.* Aurora, Ont.: Garamond Press.

Tokar, Brian
(1987) *The Green Alternative: Creating an Ecological Future.* San Pedro, Calif.: R. and E. Miles.

Turton, David
(2002) "Forced Displacement and the Nation-State," in: *Development and Displacement,* edited by Robinson. Oxford: Oxford University Press.
Vanderwerth, W. C., ed.
(1971) *Indian Oratory: Famous Speeches by Noted Indian Chieftains.* Norman: University of Oklahoma Press.

Uukw, Delgam and Gisday Wa
(1992) *The Spirit In the Land: Statements of the Gitksan and Wet'suwet'en Hereditary Chiefs in the Supreme Court of British Columbia 1987-1990.* Gabriola, British Columbia: Reflections.

Wackernagel, Mathis and William Rees
(1996) *Our Ecological Footprint: Reducing Human Impact on the Earth.* Gabriola Island, B. C.: New Society.

Waldram, James B.
(1988) *As Long as the Rivers Run: Hydroelectric Development and Native Communities in Western Canada.* Winnipeg: University of Manitoba Press.
Warren, D. M., L. J. Slikkerveer and D. Brokensha, eds.
(1995) *The Cultural Dimension of Development: Indigenous Knowledge*

Systems. London: Intermediate Technology Publications.
Warren, Karen J.
  (1990) "The Power and the Promise of Ecological Feminism." *Environmental Ethics* 12:125-46.
WCED (World Commission on Environment and Development)
  (1987) *Our Common Future*. Oxford: Oxford University Press.
Weatherford, Jack
  (1988) *Indian Givers: How the Indians of the Americas Transformed the World*. New York: Fawcett Columbine.
Weiler, Michael H.
  (1992) "Contemporary Harvesting by the Moose Factory and New Post First Nations within the Moose River Basin," Ontario. Report prepared for the Moose River/James Bay Coalition.
Weiskel, Timothy C.
  (1989) "The Ecological Lessons of the Past: An Anthropology of Environmental Decline." *The Ecologist*. 19(3):98-103.
Weltman, Eric
  (2001) "Here Comes the Sun: What Ever Happened to Solar Energy," in: *Environment 01/02.*, edited by Allen, Guilford, Conn.: McGraw-Hill/Dushkin, pp. 83-6.
Wheeler, Jordan
  (1992) "Voice," in: *Aboriginal Voices: Amerindian, Inuit and Sami Theatre*, edited by Brask and Morgan. Baltimore: The Johns Hopkins University Press, pp. 37-43.
Widdowson, Frances and Albert Howard
  (2006) "Aboriginal 'Traditional Knowledge' and Canadian Public Policy: Ten Years of Listening to the Silence," Presentation for the Annual Meeting of the Canadian Political Science Association, York University, Toronto, Ontario, June 1-3, 2006.
White, Jr., Lynn
  (1968) "The Historical Roots of Our Ecological Crisis," *Science* 162:1243-48.
Williams, Nancy M. and Graham Baines, eds.
  (1993) *Traditional Ecological Knowledge: Wisdom for Sustainability*, Canbera: Centre for Resource and Environmental Studies, Australian National University.
Williams, Ted
  (1999) "Lessons from Lake Apopka," *Audubon*, July/August, pp. 64-72.
Winter, James
  (1997) *Democracy's Oxygen: How Corporations Control the News*. Montreal: Black Rose Books.
Wolf, Eric R.
  (1982) *Europe and the People Without History*. Berkeley: University of California Press.
Wooden Leg, with Thomas B. Marquis, ed.
  (1931) *Wooden Leg: A Warrior Who Fought Custer*. Lincoln: University of

Nebraska Press.

Wright, Ronald
(1992) *Stolen Continents: The New World Through Indian Indian Eyes Since 1492,* Toronto: Viking/Penguin.

ABOUT THE AUTHOR

Dr. Roy C. Dudgeon is a radical ecologist, anthropologist and philosopher from Winnipeg, Manitoba, Canada. He received his Ph. D. in cultural anthropology from the University of Manitoba in 2003. He also holds an M. A. in social anthropology from York University in Toronto, and a B. A. (Hons.) in anthropology and philsosophy from the University of Winnipeg, where he received the Governor General's Silver Medal for the Highest Standing in Arts (Hons.) upon graduation. Some of the ideas contained in *Common Ground* were first pursued there, developed in greater detail through his Ph. D. research, then revised and published as the current book.

Dr. Dudgeon currently teaches part time at the University of Winnipeg, and is a founding member of the Green Party of Manitoba.

Also available from Pitch Black Publications:

Roy C. Dudgeon (2008) *The Pattern Which Connects: An Eco-holist Critique of Postmodernism.*

9 781435 717381